Exploring the Universe

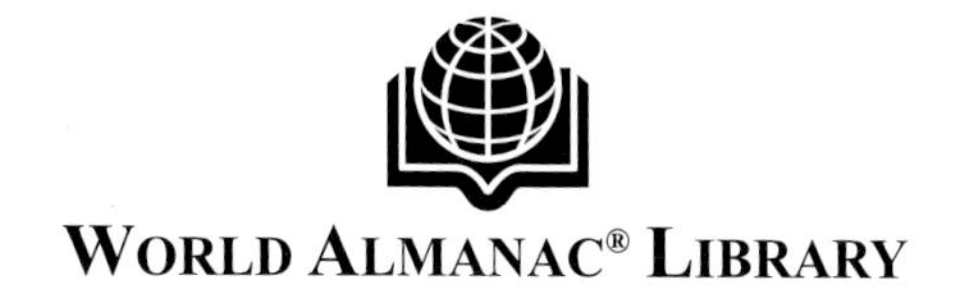

WORLD ALMANAC® LIBRARY

Please visit our website at: www.worldalmanaclibrary.com
For a free color catalog describing World Almanac® Library's list of high-quality books and multimedia programs, call 1-800-848-2928 (USA) or 1-800-461-9120 (Canada). World Almanac® Library's Fax: (414) 332-3567.

The editors at World Almanac® Library would like to thank Greg Walz-Chojnacki, a writer, editor, and consultant in the study of astronomy and space technology, for the technical expertise and advice he brought to the production of this book.

Library of Congress Cataloging-in-Publication Data

Exploring the universe. — North American ed.
p. cm. — (21st century science)
Includes bibliographical references and index.
ISBN 0-8368-5001-7 (lib. bdg.)
1. Astronomy—Juvenile literature. 2. Cosmology—Juvenile literature. [1. Astronomy. 2. Universe.] I. Title. II. Series.
QB46.E96 2001
520—dc21 2001031442

This North American edition first published in 2001 by
World Almanac® Library
330 West Olive Street, Suite 100
Milwaukee, WI 53212 USA

Created and produced as the *Visual Guide to the Wonders of Our Universe* by
QA INTERNATIONAL

329 rue de la Commune Ouest, 3e étage
Montreal, Québec
Canada H2Y 2E1
Tel: (514) 499-3000 Fax: (514) 499-3010
www.qa-international.com

Editorial Director: François Fortin
Executive Editor: Serge D'Amico
Art Director: Marc Lalumière
Graphic Designer: Anne Tremblay
Writers: Nathalie Fredette, Claude Lafleur
Computer Graphic Artists: Mamadou Togola, Alain Lemire, Hoang-Khanh Le, Ara Yazedjian, Mélanie Boivin, Jean-Yves Ahern, Michel Rouleau
Page Layout: Lucie Mc Brearty, Véronique Boisvert, Geneviève Théroux Béliveau
Researchers: Anne-Marie Villeneuve, Anne-Marie Brault
Astronomy Reviewer: Louie Bernstein
Copy Editor: Jane Broderick
Production: Gaétan Forcillo, Guylaine Houle
Prepress: Tony O'Riley
Translation: Kathe Roth
World Almanac® Library Editor: David K. Wright
World Almanac® Library Art Direction: Karen Knutson
Cover Design: Katherine A. Kroll

Photo credits: abbreviations: t = top, c = center, b = bottom, r = right, l = left
p. 6 (NGC 1232 & 1365): © ESO; p. 6 (Great Magellanic Cloud): Association of Universities for Research in Astronomy Inc. (AURA), all rights reserved; p. 6 (Sombrero galaxy): Association of Universities for Research in Astronomy Inc. (AURA), all rights reserved/NSF; p. 9 (tr): Lund Observatory, Sweden; p. 10 (bl): Jason Ware/The Electronic Universe Project/NASA; p. 12 (bl): ESO NTT and Herman-Josef Roeser/HST/NASA; p. 13 (tr): AURA/STScI; p. 13 (br): NRAO/AUI; p. 21 (br): COBE Science Team/DMR/NASA; pp. 24–25 (from l to r): NASA/Compton Observatory Egret Team, ROSAT All-Sky Survey, J. Bonnell and M. Perez (GSFC)/NASA, Lund Observatory, Sweden, GSFC/NASA, COBE Science Team/DMR/NASA, C. Haslam and al. (MPIfR), Skyview/NASA; p. 32 (tl): IVV/NASA; p. 33 (Cartwheel galaxy): Kirk Borne (STScI)/NASA; p. 33 (Gaseous Pillars): J. Hester and P. Scowen (ASU)/NASA; p. 33 (Hubble Deepfield): Robert Williams (STScI)/NASA; pp. 33 (Eta carinae), 42 (cr), 46 (bc & br), 47 (cl), 48 (Europa, Ganymede, & Callisto), 54 (cr): JPL/NASA; p. 34 (tl & bl): NRAO; p. 34 (br): National Astronomy and Ionosphere Center/Cornell University/NSF/NASA; p. 37 (cr): Calvin Hamilton/LPI/NASA); p. 37 (bl): Photo Researchers/NASA; p. 38 (Orion): C. R. O'Dell and S. K. Wong (Rice University)/NASA; p. 38 (Protoplanetary disks): M. J. McCaughrean (Max-Planck Institute for Astronomy)/ C. R. O'Dell (Rice University)/NASA; p. 38 (Protoplanet in Taurus): S. Tereby (Extrasolar Research Corp)/NASA; pp. 42 (cc & b), 46 (bl) U.S. Geological Survey/NASA, 48 (Io), 51 (b): U.S. Geological Survey/NASA; pp. 45 (bc & br), p. 47 (tl), 52 (bl): NSSDC/NASA; p. 48 (Jupiter): USGS/NASA; p. 52 (br): Malin Space Science Systems/NASA; p. 54 (bl): JHUAPL/NASA

Printed in Canada

1 2 3 4 5 6 7 8 9 05 04 03 02 01

Table of Contents

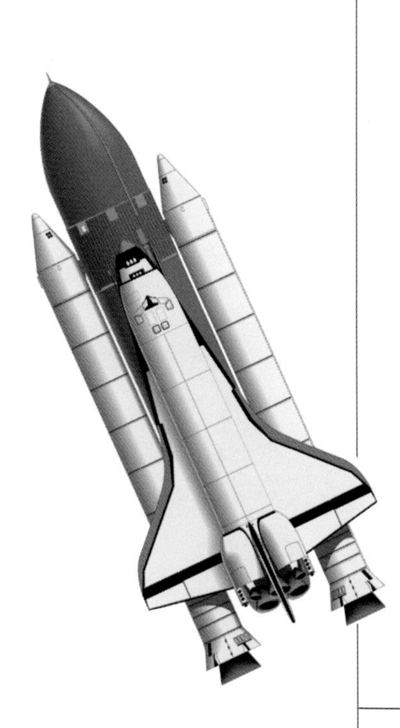

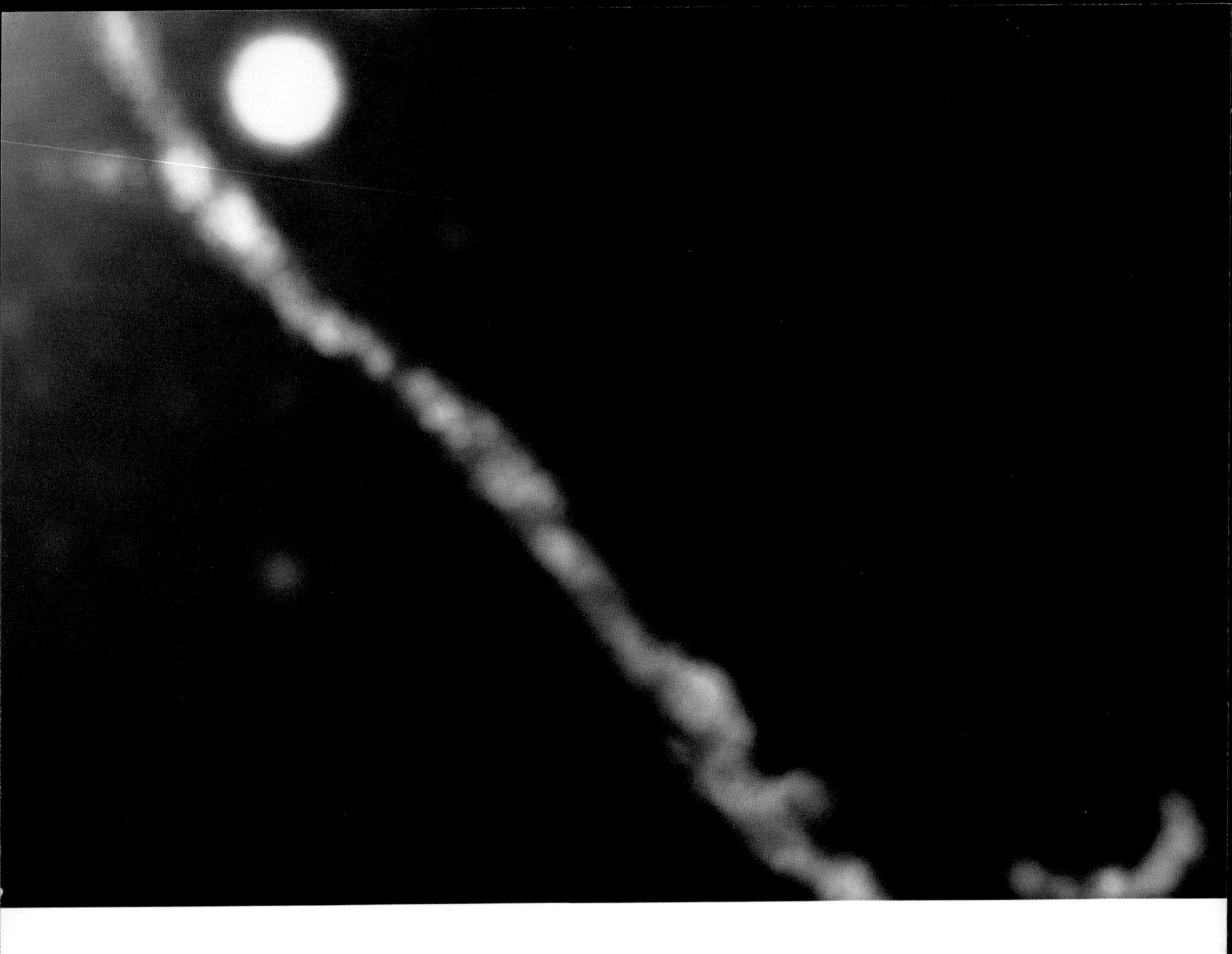

The Milky Way, the spiral galaxy in which our Sun is located, shares the Universe with billions of other galaxies, which are huge structures containing billions of stars. Surrounded by immense stretches of empty space, these galaxies form the background of the Universe. This survey of star clusters reminds us of where we fit in the grand scheme of things!

The Galaxies

The Galaxies

Billions of islands of billions of stars ...

A galaxy consists of billions of stars and interstellar dust and gas held together by gravity. Each galaxy forms a bright island lost in the immense black sea of the Universe. It is estimated that the Universe contains 100 billion galaxies and that each one contains an average of 100 billion stars. Dwarf galaxies contain only a few million, while giant galaxies contain trillions. The diameter of galaxies ranges from about 3,000 to more than 500,000 light-years. A light-year is the distance light travels in one year — about 5.878 trillion miles (9.46 trillion kilometers).

THE BIRTH OF A GALAXY

About 2 billion years after the Big Bang, galaxies apparently formed from diffuse clouds of gas and primordial material.

Gravity began to pull the material toward the center. As it collapsed, the cloud slowly began to rotate.

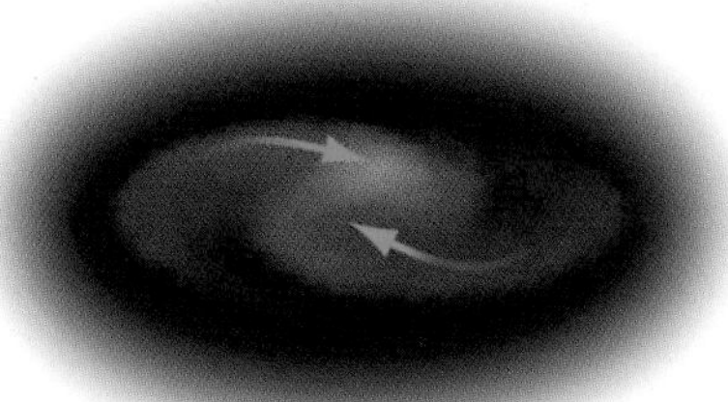

The cloud then flattened to form a disk with a large central bulge in which new stars were born.

Over time, the disk flattened even more, and eventually spiral arms formed.

GALAXIES OF ALL SIZES AND SHAPES

In the constellation Eridanus is a splendid spiral galaxy, **NGC 1232**. Young stars can be seen throughout its long arms.

The **Sombrero** galaxy, in Virgo, is a good example of a hybrid lenticular spiral galaxy featuring an enormous nucleus.

NGC 1365 is a barred spiral galaxy in the constellation Fornax, fifty light-years from Earth.

The **Large Magellanic Cloud** is a typical irregular galaxy, located close to our galaxy, the Milky Way.

Classification of the Galaxies

Telling one island from another

In 1925, the astronomer Edwin Hubble designed a simple method for classifying galaxies that is still in use today. He identified three main shapes — elliptical, spiral, and irregular — and later added a fourth, lenticular. About 60 percent of observed galaxies are spiral in shape, 20 percent are lenticular, 15 percent are elliptical, and between 3 percent and 5 percent are irregular.

ELLIPTICAL GALAXIES

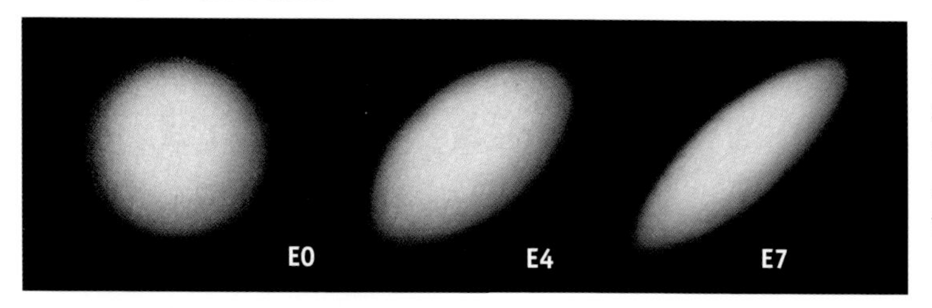

Elliptical (E) galaxies have spheroidal shapes that smooth down toward the edges. They are classified from 0 to 7 according to the length of the ellipse.

LENTICULAR GALAXIES

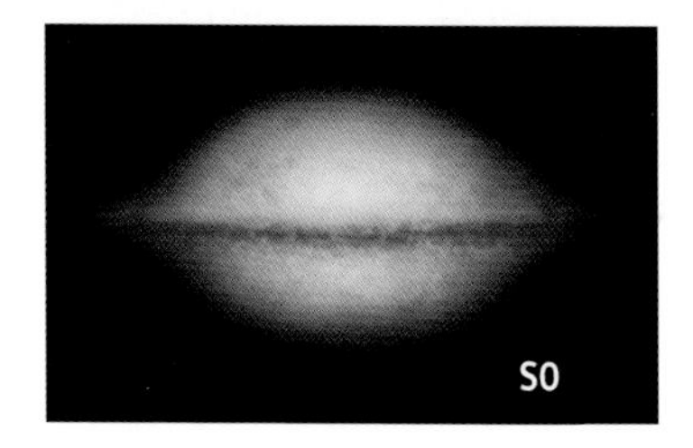

Lenticular (S0) galaxies resemble very flattened elliptical galaxies, but they have a central bulge similar to those of spiral galaxies.

SPIRAL GALAXIES

❶

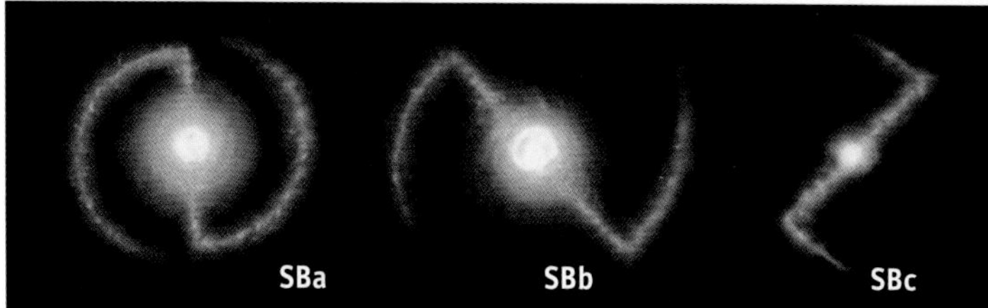

❷

Spiral galaxies have curved, spiral-shaped arms on the sides of their nucleus. They are classified Sa, Sb, and Sc according to the size of the nucleus and how tightly the spiral arms are wound. The Milky Way is an Sb-type spiral galaxy.

Normal spiral galaxies (S) ❶ often have two arms emerging from opposite sides of the nucleus. Barred spiral galaxies (SB) ❷ are crossed by a band of stars and interstellar material, from the ends of which the spiral arms emerge.

IRREGULAR GALAXIES

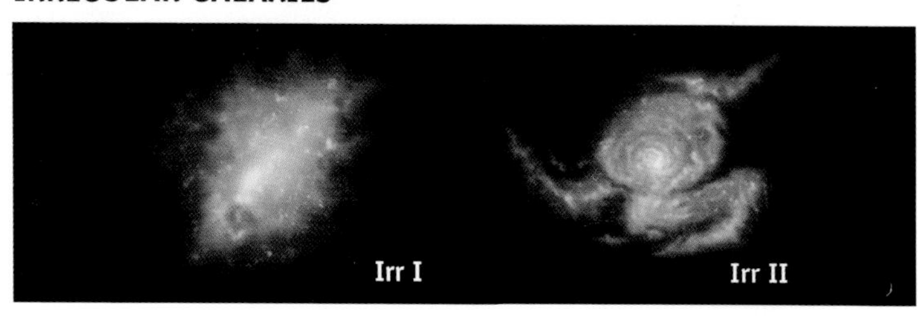

Irregular galaxies do not have a nucleus, arms, or any particular shape. Type I (Irr I) irregular galaxies have no defined structure, while type II (Irr II) irregular galaxies present a perturbed, or distorted, structure.

The Milky Way

Our island in the Universe

Our Solar System is in a galaxy called the Milky Way. Seen from Earth, it forms a narrow, dim cloud crossing the nocturnal sky from one side to the other. It looks a little like a stream of milk, which is what inspired the ancient Greeks to name it the Milky Way.

Our galaxy, composed of over 100 billion stars forming a large disk with spiral arms, is 10 billion years old, while the Solar System is about 5 billion years old.

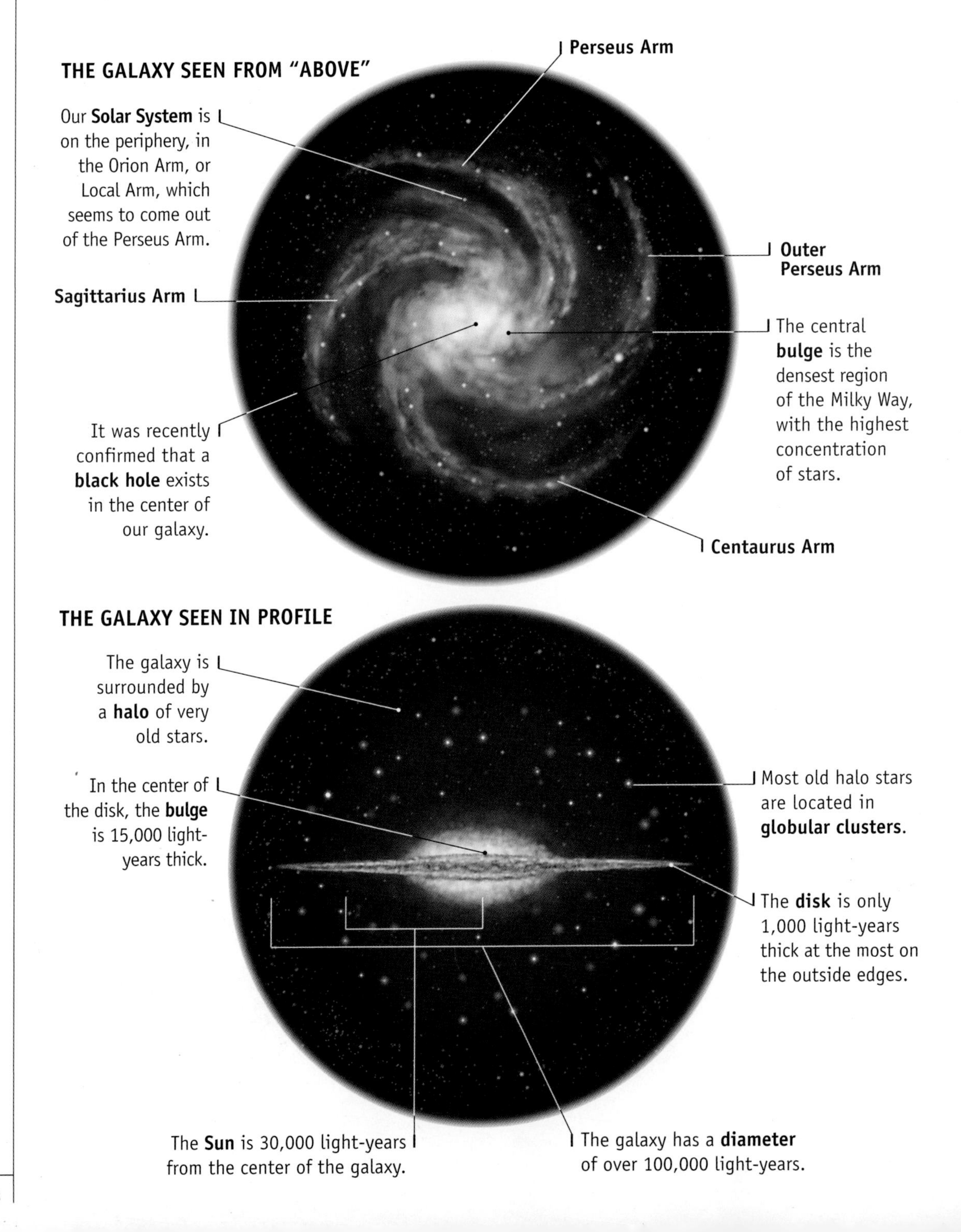

THE PANORAMA OF THE MILKY WAY

It is difficult to determine the exact shape of our galaxy, because the fact that we are within it means that we cannot see the whole thing. We are slightly above the galactic equator, and the nucleus appears to us to be in the direction of Sagittarius. Unfortunately, we cannot see the bulge, because it is hidden behind dense dust.

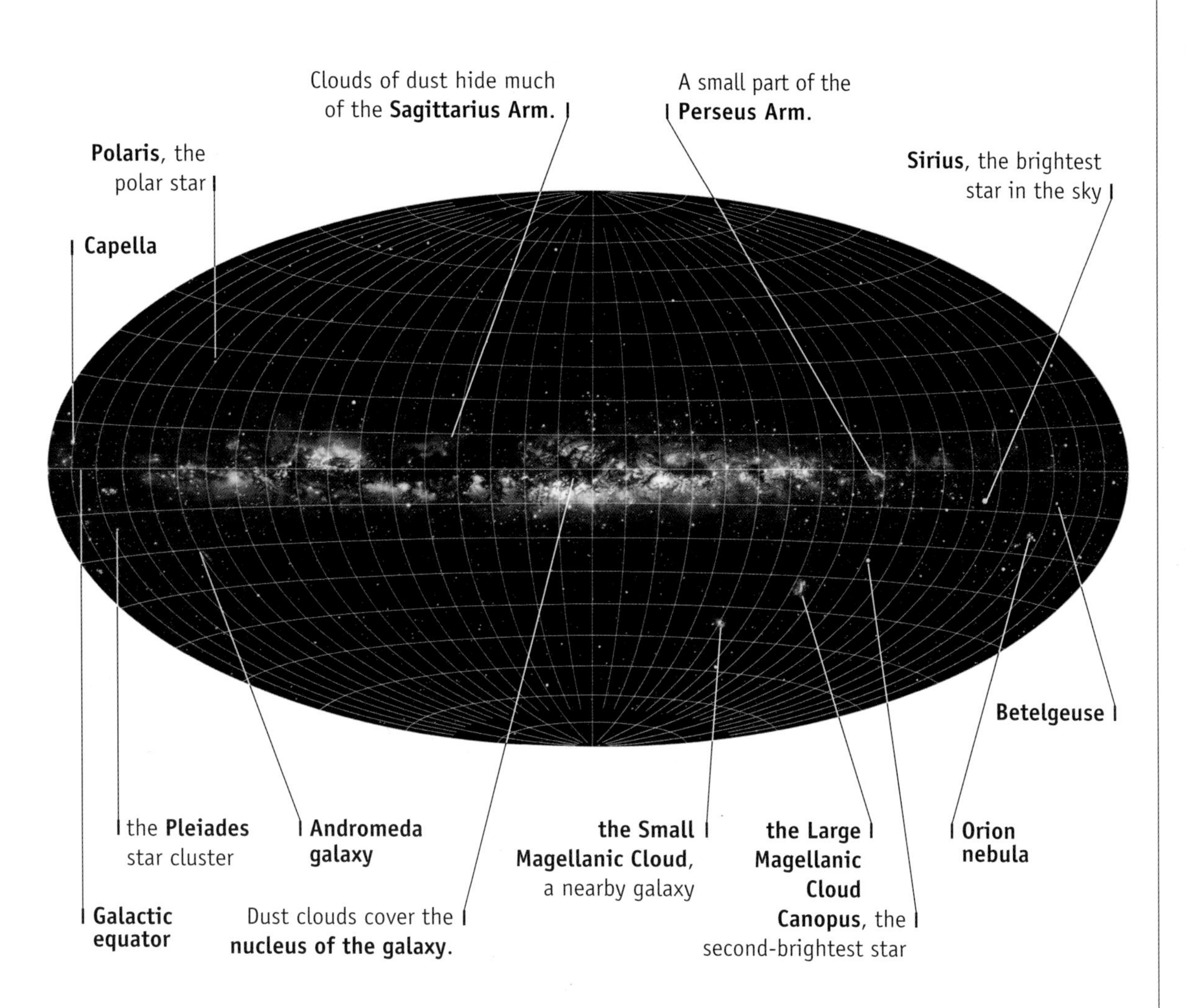

REVOLVING AROUND THE GALACTIC NUCLEUS

Earth rotates once every twenty-four hours, at a speed of 1,040 miles (1,670 km) per hour ❶. Earth makes one revolution around the Sun each year, at a speed of 66,450 miles (107,000 km) per hour, traveling 1.55 million miles (2.5 million km) per day ❷. The Sun, for its part, revolves around the galactic nucleus at 620,000 miles (1 million km) per hour, taking 220 million years to make a complete revolution ❸. Since it came into existence, the Solar System has made only twenty revolutions of the Milky Way.

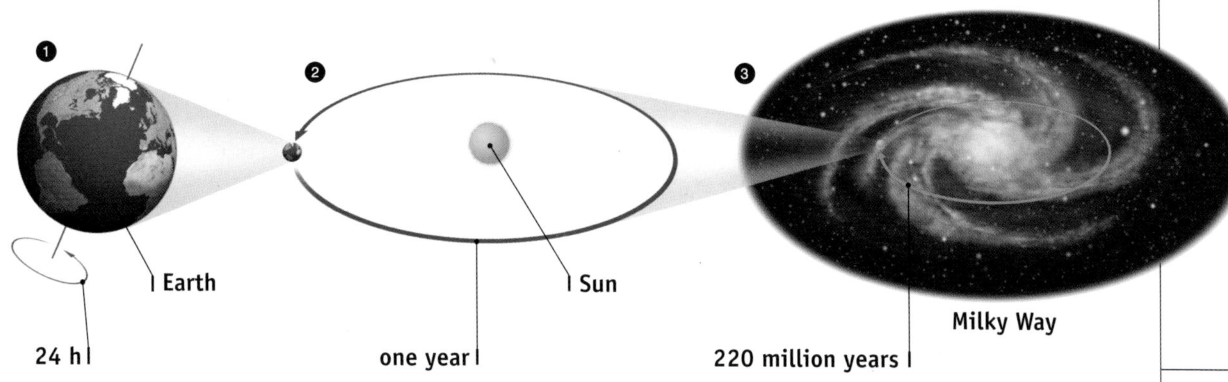

The Local Group

Our neighboring galaxies

Like stars, galaxies have a tendency to group together. The Milky Way, our galaxy, is part of a local group (or cluster) that includes some thirty galaxies. Our galaxy and the Andromeda galaxy are the largest members of the group; most of the others are small elliptical or irregular galaxies. The entire local group extends over about 6 million light-years.

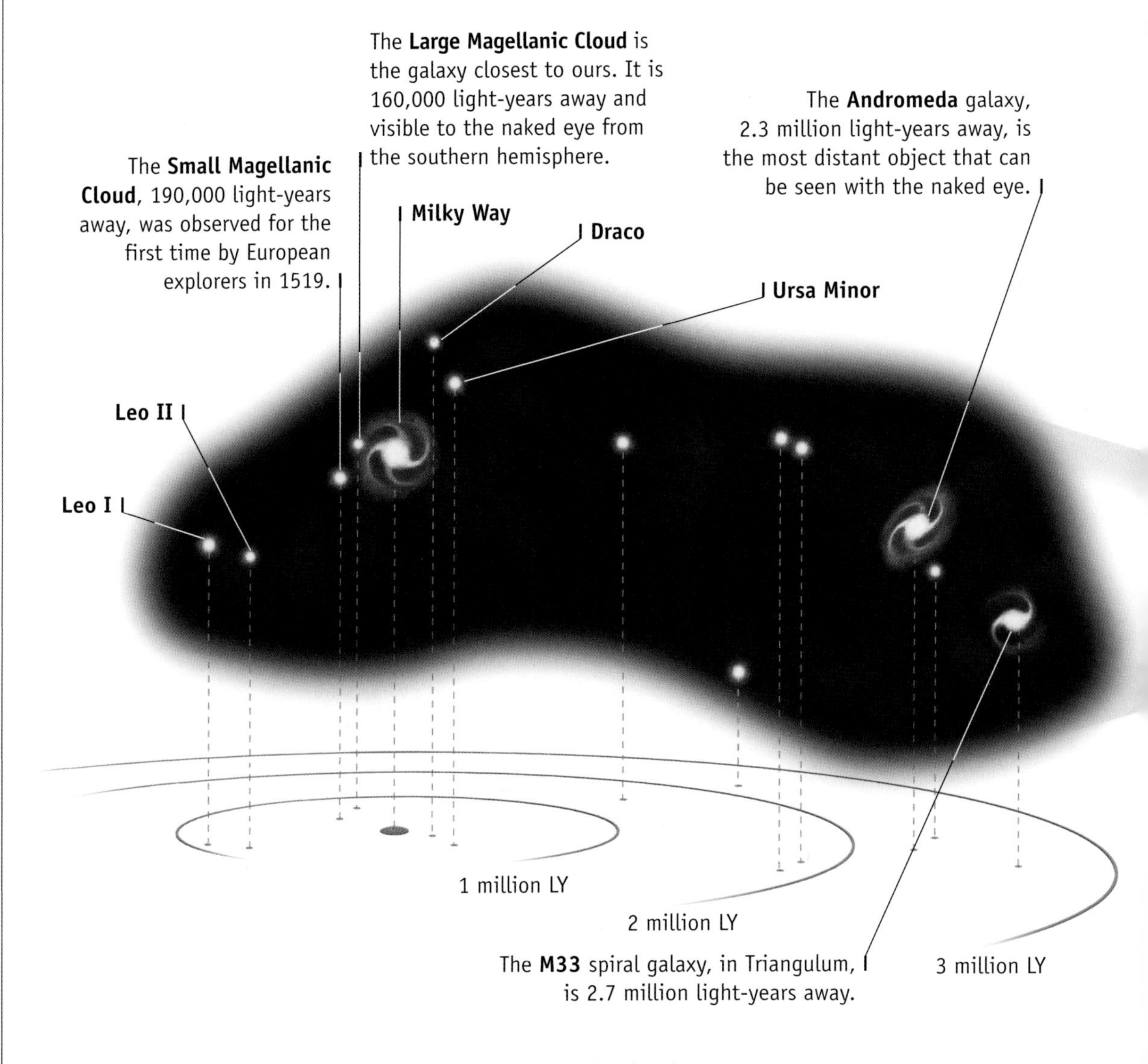

The **Andromeda** Galaxy is a spiral galaxy that is very similar to the Milky Way. It is slowly getting closer to our galaxy and should collide with it in 10 billion years.

Galactic Clusters

Huge groupings in the Universe

Galactic groups, or clusters, contain between a few and thousands of galaxies. "Rich" clusters are major concentrations of large galaxies, generally gathered in a defined structure (spherical or ellipsoidal in shape). There are also "poor" clusters, which have an irregular shape and contain fewer galaxies.

THE LOCAL SUPERCLUSTER

This colossal grouping extends over more than 100 million light-years and includes many clusters and thousands of galaxies. The local supercluster is far from unique; some fifty similar groupings have been found, containing an average of twelve rich clusters. Some astronomers are now searching for even larger structures.

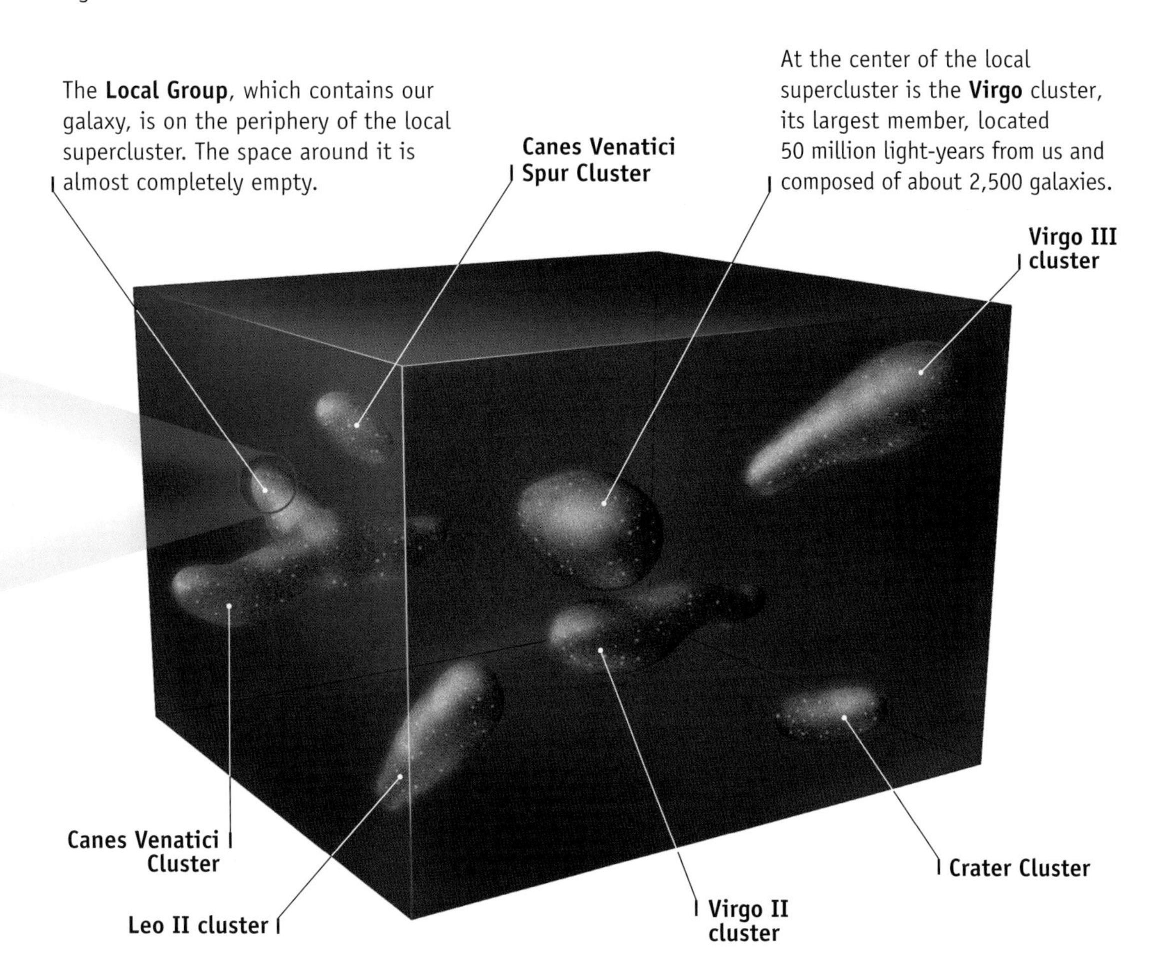

LIKE A SPONGE

Astronomers have observed large networks of superclusters spread over hundreds of millions of light-years. Clusters and superclusters are separated by "bubbles" that are almost empty of galaxies; some bubbles are more than 300 million light-years in diameter. Thus, we can think of the Universe as looking something like a sponge!

Active Galaxies

Intense energy in the heart of galaxies

There is a very specific and widely varied family of galaxies called active galaxies. These galaxies emit much of their radiation in the form of X rays, infrared rays, and radio waves.

Active galaxies (which include radio galaxies, Seyfert galaxies, and quasars) always have a distinct shape, distorted sometimes by the presence of neighboring galaxies. They emit large quantities of energy, generally more than that from ordinary galaxies, mainly in the form of visible light. It is thought that active galaxies are fed by black holes in their center.

QUASARS

The most unusual type of active galaxy is the quasar (contraction of "quasi-stellar radio source"). Discovered in the 1960s, these objects are probably the size of our Solar System and emit more energy than a galaxy composed of hundreds of billions of stars. Quasars are the most distant objects in the Universe that we can observe; their light was emitted billions of years ago.

3C 273 Quasar was one of the first to be discovered.

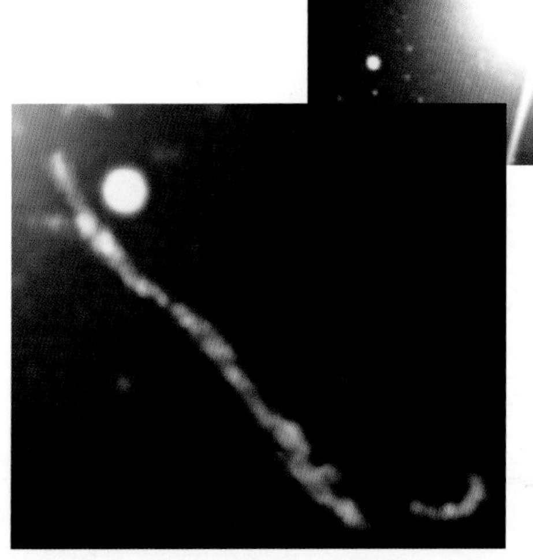

This photograph taken by the Hubble Space Telescope shows a jet coming from the nucleus of the quasar.

Colossal forces at the center of the galaxy give rise to formidable jets of **matter.**

SEYFERT GALAXIES

In 1943, the astronomer Carl Seyfert discovered a type of galaxy whose nucleus is particularly bright. There are about 150 Seyfert galaxies, most of which are normal spirals that emit large amounts of infrared rays and few radio waves.

Seyfert galaxy **NGC 7742** resembles a normal spiral galaxy, but its nucleus is very bright.

RADIO GALAXIES

Radio galaxies are giant elliptical galaxies that can emit up to 100,000 times more radio waves than an ordinary galaxy. The radio waves may come from the center of the galaxy, originating in a region that can be either very small or very large. Radio galaxies often contain two regions emitting radio waves; these regions may be separated from each other by millions of light-years.

The exceptional luminosity of quasars is thought to be due to the presence of a **black hole**, which swallows up the material in the vicinity and creates an energy jet.

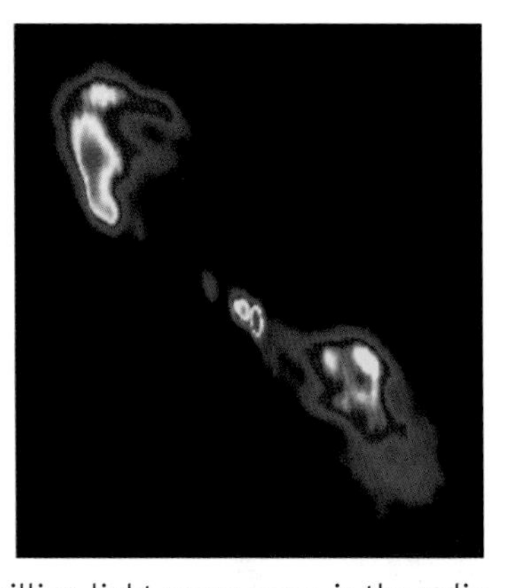

Centaurus A, a galaxy 15 million light-years away, is the radio galaxy closest to us. Left, the galaxy in visible light, crossed by a large band of dust. Right, the radio image shows two lobes on either side of the galaxy, at a 90° angle to the band. The invisible radio-wave emissions come from these lobes, which are almost 2 million light-years in size.

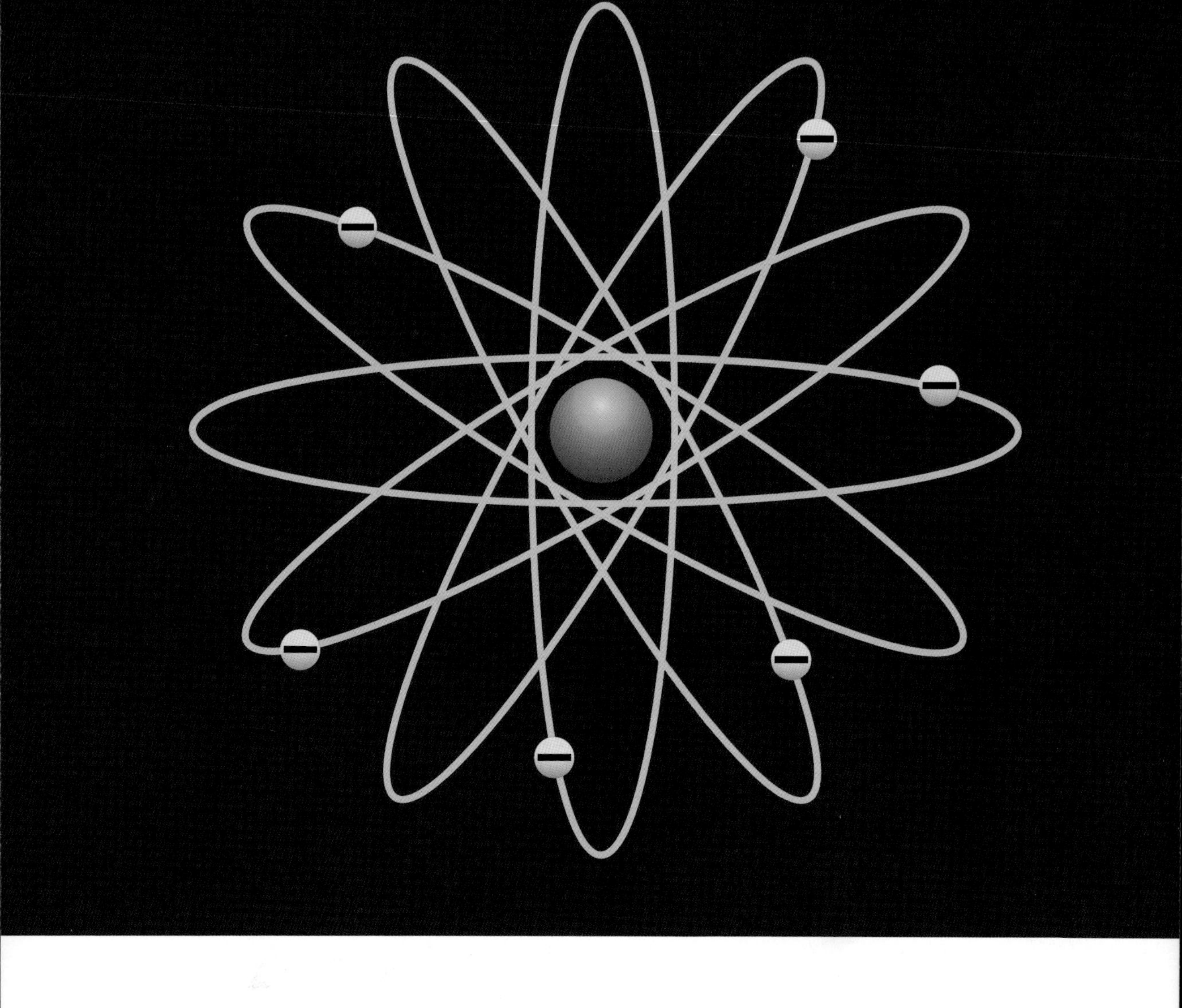

The fact that the galaxies are moving away from each other suggests that the Universe is expanding. If this is true, then how long, we may wonder, has this been going on? How did the Universe start? What is the Big Bang? Can we still find traces of this original cosmic explosion? What will be the fate of the Universe? This section offers some answers to these awe-inspiring questions.

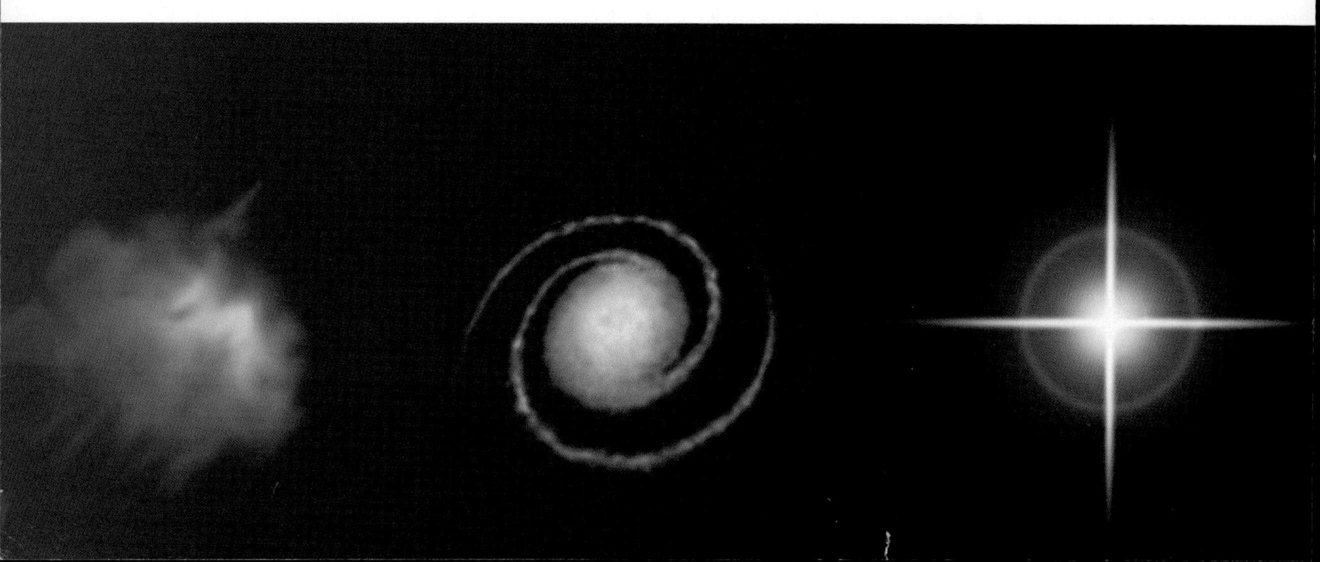

Structure of the Universe

The Size of the Universe

From infinitely small to infinitely large

Normally, Earth seems immense to us. On the scale of the Universe, however, it is tiny, especially when we consider that distances in the Universe can easily be measured in trillions of miles or trillions of kilometers — or, more conveniently, in light-years. Our Solar System is itself part of a galaxy, one of 100 billion galaxies that make up the Universe.

THE UNIVERSE SEEN AS NESTING DOLLS

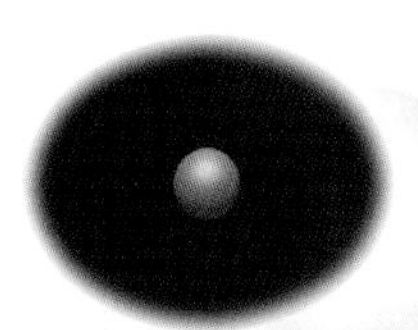

38^{-18} foot (10^{-18} meter)

Despite its variety of shapes and forms, matter is made up of a small number of simple constituents. The smallest particle of matter is a **quark**.

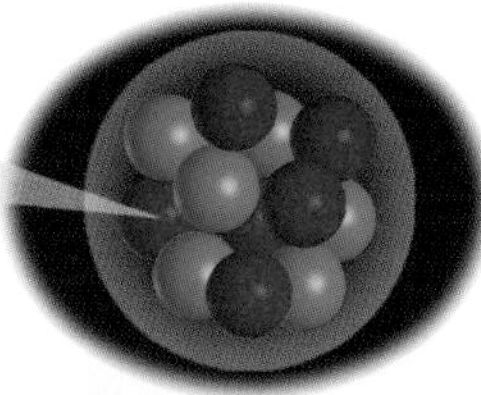

38^{-15} ft (10^{-15} m)

Quarks group together to form protons and neutrons, forming the basis for an **atomic nucleus**.

38^{-10} ft (10^{-10} m)

This nucleus is in the center of an **atom**.

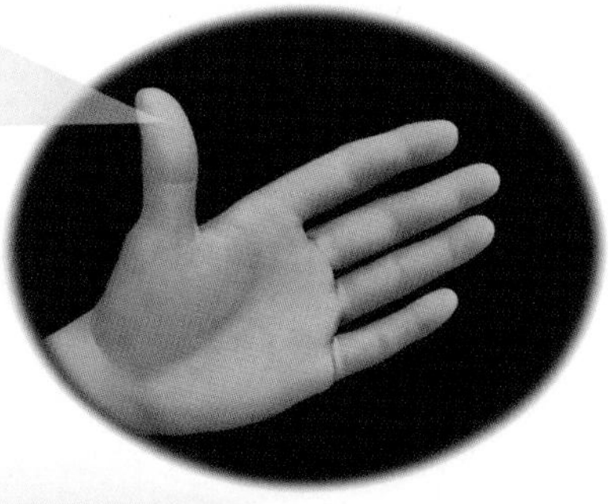

38^{-1} ft (10^{-1} m)

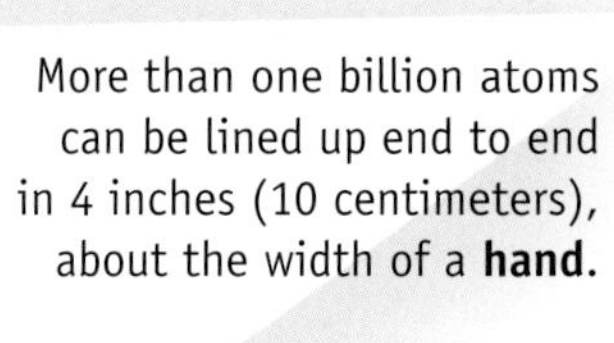

More than one billion atoms can be lined up end to end in 4 inches (10 centimeters), about the width of a **hand**.

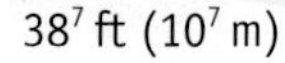

38^{7} ft (10^{7} m)

Our planet, **Earth**, has a diameter of 7,920 miles (12,756 km).

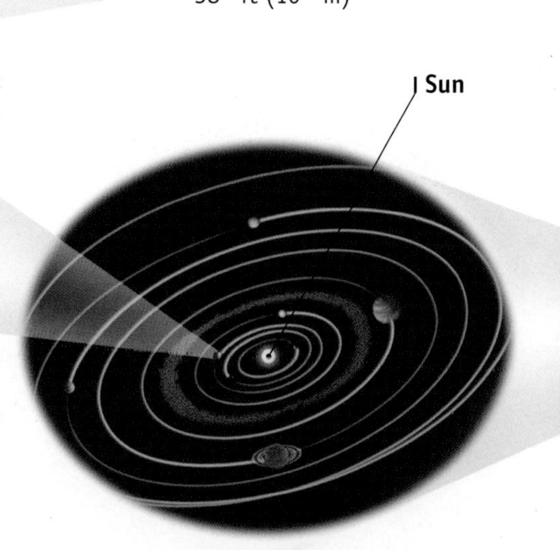

38^{13} ft (10^{13} m)

The **Solar System** is composed of nine planets and one star, our Sun. The entire system, stretching more than 7.5 billion miles (12 billion km), is located in one of the spiral arms of our galaxy.

Astronomers created new units with which to measure the Universe, taking the distance from Earth to the Sun as the standard unit, or astronomical unit (AU). This is the average distance between us and our star. Longer distances are expressed in another unit: the light-year (LY), which is the distance that light travels in one year at a speed of 186,300 miles (300,000 km) per second — about 5.878 trillion* miles (9.46 trillion km) per year.

MEASUREMENT UNITS	
unit	**value**
astronomical unit (AU)	93 million miles (149.6 million km)
light-year (LY)	5,878 billion* miles (9,460 billion km)
parsec (pc)	3.26 light-years or 206,265 AU
megaparsec (Mpc)	3,260,000 light-years

*One trillion equals 1,000 billion, or 1,000,000,000,000.

Here's one way to think of the difference between one thousand, one million, and one billion: one thousand seconds are equivalent to about fifteen minutes; one million seconds are equivalent to almost two weeks; and one billion seconds are equivalent to thirty-two years.

Galactic superclusters form the complex canvas of the **Universe**, which contains about 100 billion galaxies. In this almost unimaginable structure is a tangled network of clusters, galactic superclusters, and immense bubbles of emptiness.

38^{30} ft (10^{30} m)

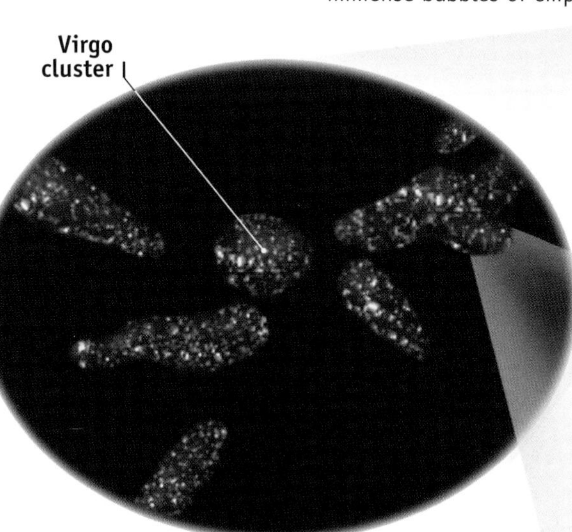

38^{23} ft (10^{23} m)

The Local Group is itself part of a galactic **supercluster**, a complex, filamentous structure that extends for 100 million light-years.

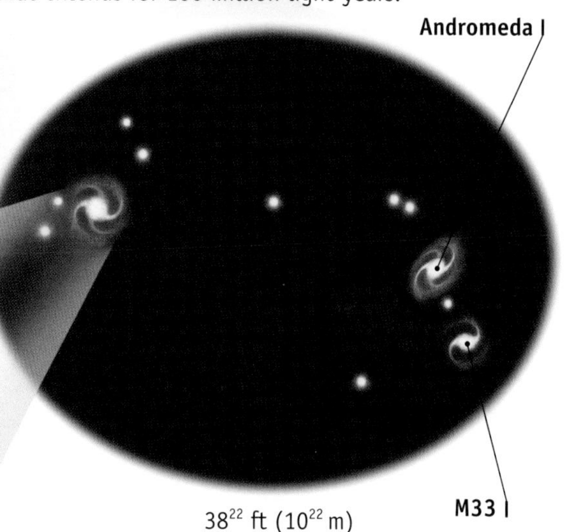

38^{22} ft (10^{22} m)

The Milky Way is part of a group of thirty galaxies, the **Local Group**, that are "squeezed" together in a space of 10 million light-years.

38^{20} ft (10^{20} m)

Our galaxy, the **Milky Way**, contains more than 100 billion stars. This wide, very flat cloud measures one hundred light-years in diameter and is about ten light-years thick.

The Big Bang

The first moments of the Universe

Before the Big Bang, there was nothing, absolutely nothing — no matter, no energy, no gravitational force, not even time. Then, most scientists agree, suddenly, about 15 billion years ago, the Big Bang took place: this incredibly huge explosion gave birth to the Universe. Although it is a difficult idea to grasp, the Big Bang marked the beginning of space, matter, and time. As for what time zero and what went before it were like, science cannot yet say; it is an enigma.

During the first fractions of the first second, only energy existed. Under the impulse of the explosion, this energy spread and cooled; it became matter, which began to organize in more and more complex ways. The Universe began to expand, and it is still doing so today.

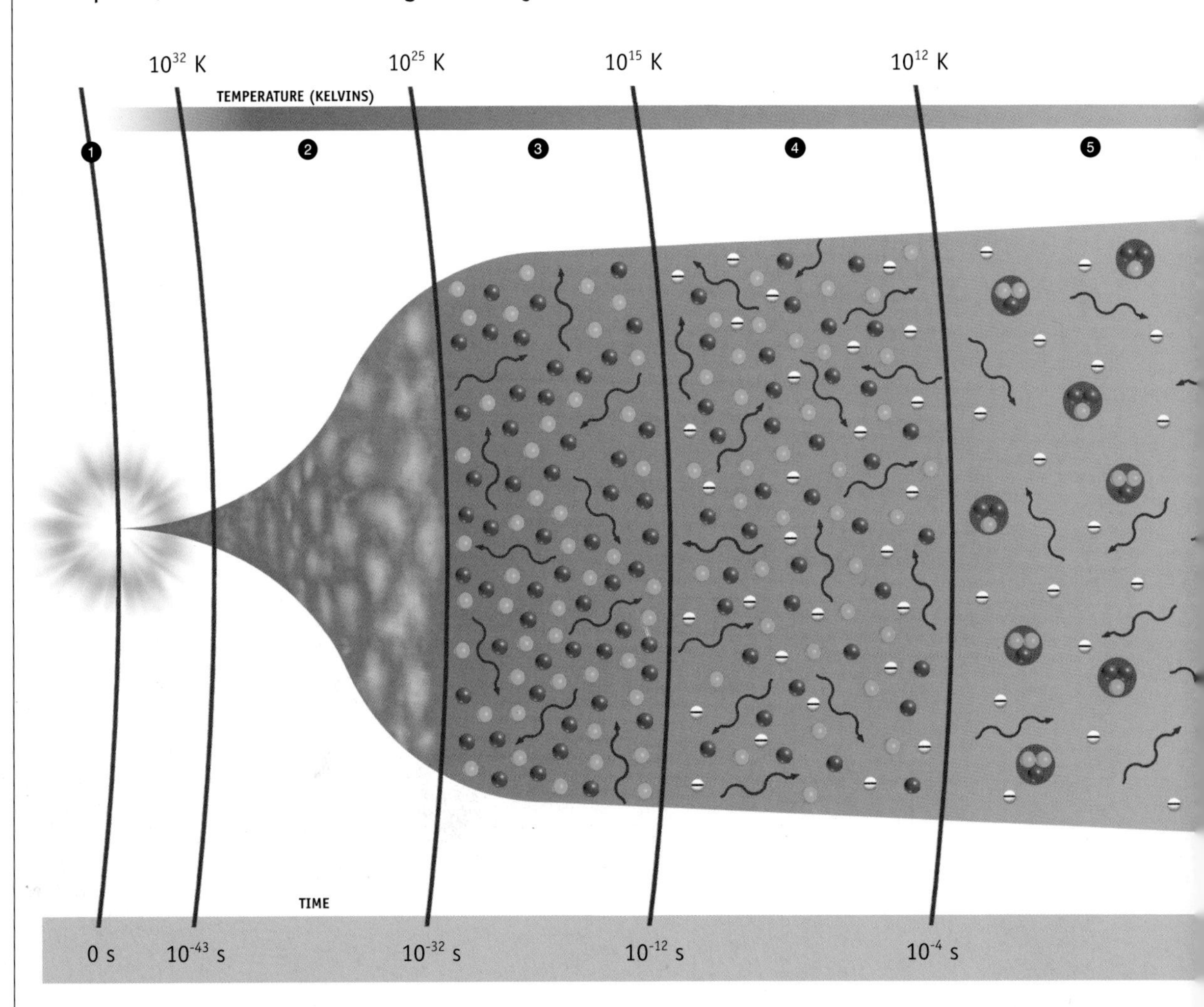

10-43 SECONDS FROM THE BEGINNING

The Big Bang is the most widely accepted theory in the scientific community to explain the birth of the Universe. By applying known laws of physics and taking advantage of the huge progress made in astronomical observation, scientists are trying to travel back up the road toward the early Universe. They have succeeded in returning to a tiny portion of the first second of the Universe; written out in full, this fraction equals 0.0000000000000000000000000000000000000000001 second after the Big Bang.

FROM THE FIRST SECONDS . . . TO TODAY

At 0 seconds, an infinitely hot and dense state concentrated all the mass of the Universe into a tiny physical point ❶. Unmeasurable amounts of energy were liberated and the original singularity expanded ❷. The initial energy was transformed into matter; elementary particles such as photons and quarks formed ❸. Gradually, the Universe cooled and continued to expand. Other particles formed, including electrons ❹. Soon after, quarks grouped together to form protons and neutrons, the basic constituents of future atomic nuclei ❺. After three minutes, the temperature had dropped, and protons and neutrons grouped together to form the nuclei of the first light elements in the Universe: hydrogen and helium ❻. When the temperature dropped to less than 3,000° K, after 300,000 years, electrons joined with protons to form the first stable hydrogen and helium atoms, and later the first molecules formed ❼. After 2 billion years, the effect of gravity caused nebulae, embryonic galaxies (or protogalaxies), galaxies, and the first stars to be formed, as matter agglomerated in space ❽. More than 8 billion years later, the Sun and the planets of the Solar System were formed ❾. Then, molecules formed more complex entities, leading to the appearance of life ❿.

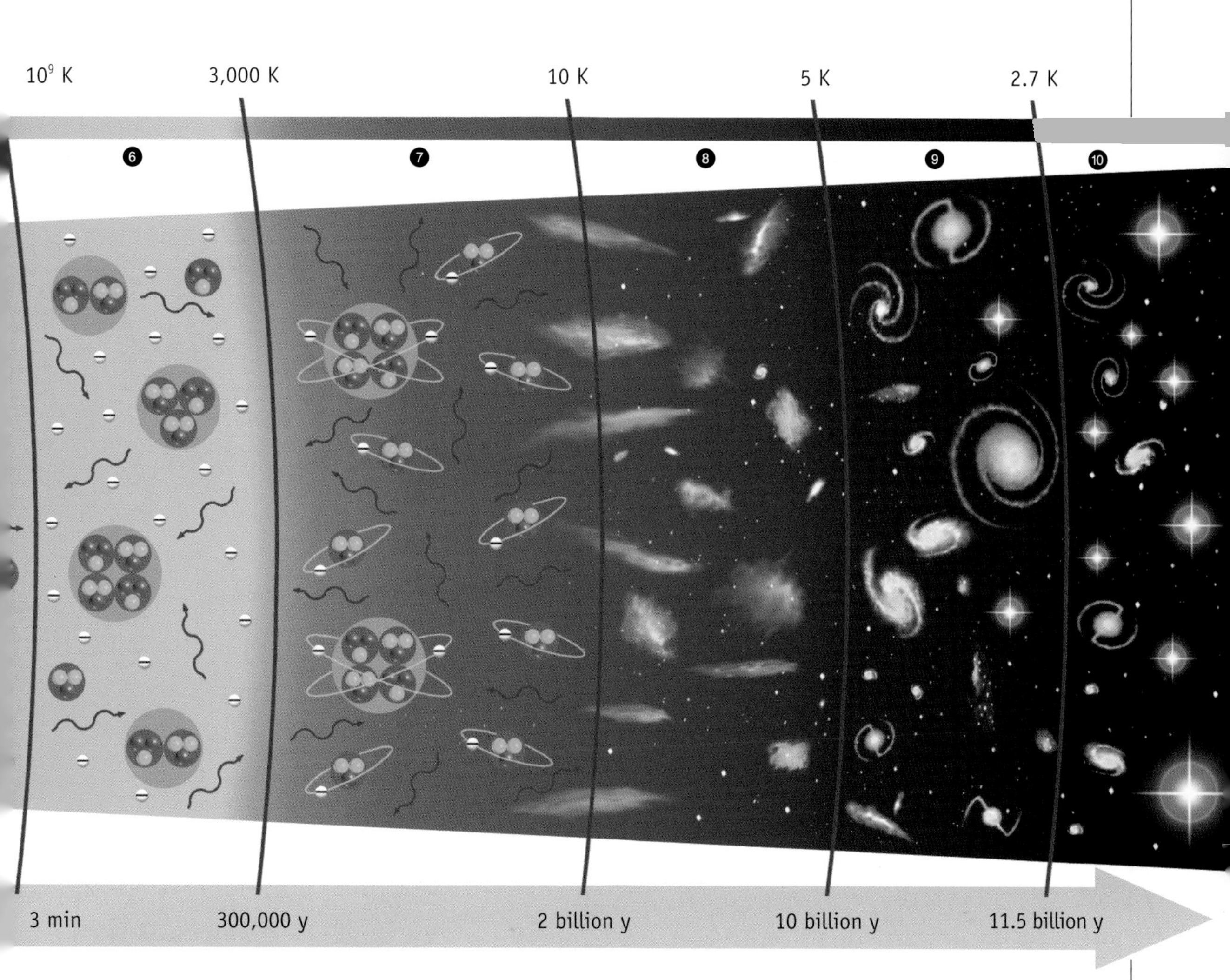

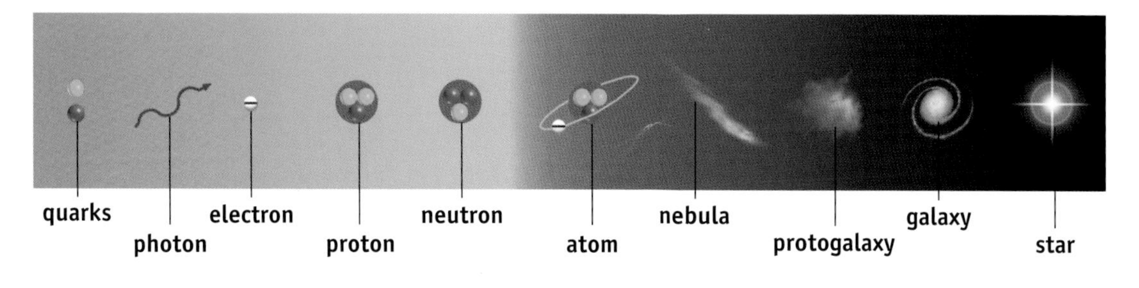

The Universe Expands

The fate of billions of galaxies

In the early 1920s, the Universe was thought to consist of our galaxy and a few objects close to it and to measure no more than 200,000 light-years. This conception changed when astronomer Edwin Hubble measured the distance to the Andromeda galaxy, which resembles ours. After that, many more distant galaxies were discovered; today, we think the Universe is made of about 100 billion galaxies.

HUBBLE'S LAW

Hubble observed that galaxies move away from each other more rapidly the more distant they are. In 1929, he formulated a law stating that the speed at which galaxies move apart increases as a function of distance. A simple analogy shows how this phenomenon works: imagine a sphere containing galaxies.

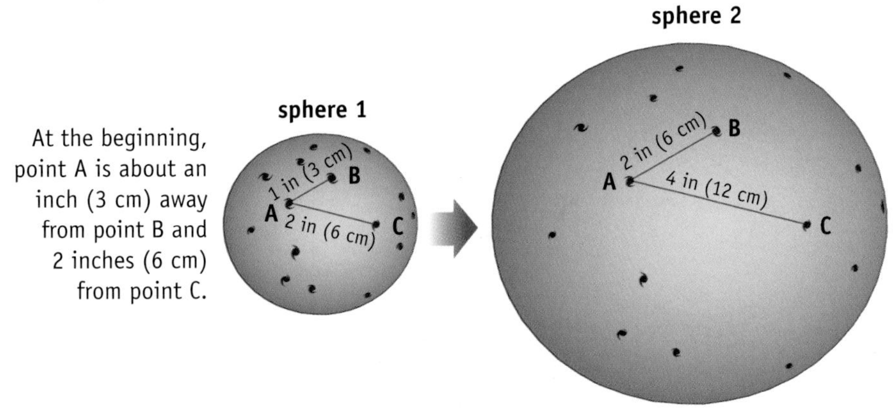

At the beginning, point A is about an inch (3 cm) away from point B and 2 inches (6 cm) from point C.

If we double the diameter of the sphere, point C moves away from point A by 2 inches (6 cm), while point B, which is closer, moves away by only 1 inch (3 cm) in the same time interval.

THE FATE OF THE UNIVERSE

As we observe it today, the Universe is expanding, but we do not know if it will always continue to do so. One of the great challenges of modern cosmology is to evaluate accurately how much matter the Universe contains, since this will determine how it will act in the future.

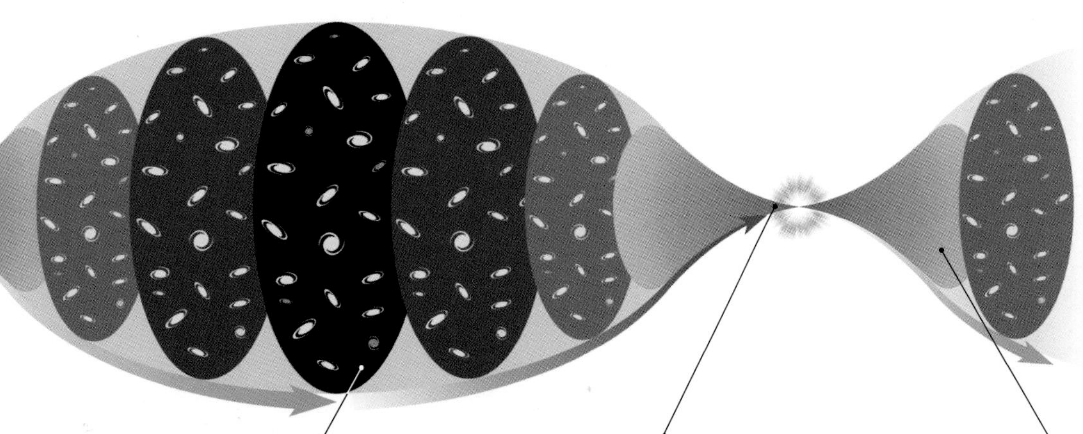

If the Universe is made up of a relatively small quantity of matter, the expansion we observe today will continue indefinitely and the Universe will expand forever. This would be an **open Universe**.

On the other hand, if the Universe is made up of a large quantity of matter, gravity will eventually stop the expansion and the Universe will contract until there is a Big Crunch. This would be a **closed Universe.**

There is speculation that the concentration of matter created by a Big Crunch would lead to a new Big Bang. In this case, it would be an **oscillating Universe**, the result of a closed Universe of Big Bangs following Big Crunches.

Cosmic Background Radiation

A voyage to the beginning of time

The more distant the star, the longer its light takes to reach us. If we look at an object 2 million light-years away, such as the Andromeda galaxy, what we are seeing corresponds to the state of this galaxy 2 million years ago, since the light it emitted then has taken that long to reach us. Looking far into the cosmos means looking back in time; the farther we look, the younger the Universe we see.

Even today, the Universe contains traces of the heat generated during the tremendous Big Bang. This residual heat is called cosmic background radiation. Whichever direction we look, we measure a uniform temperature, 2.7° above absolute zero (-459.67° F [–273° C]).

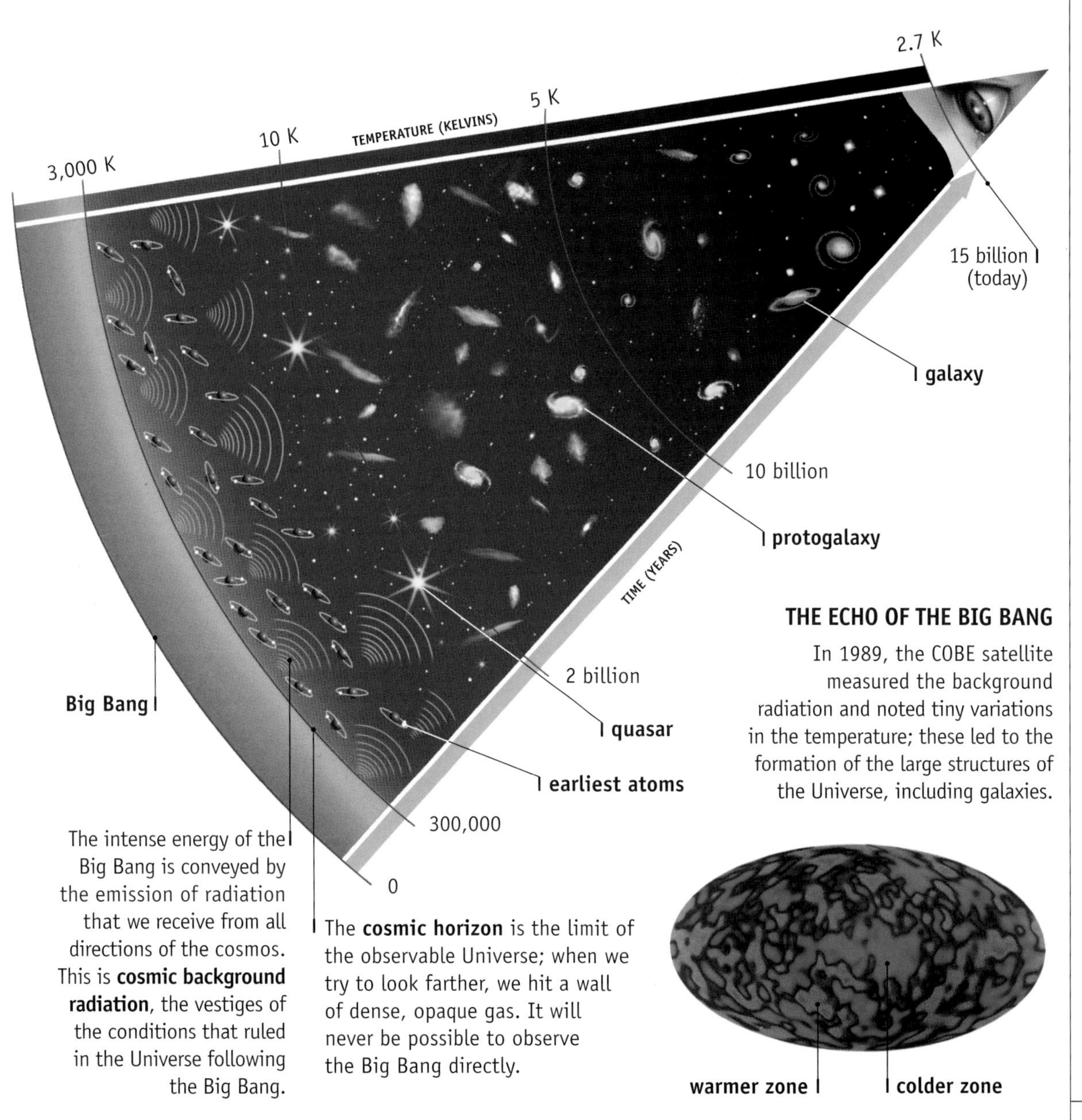

The intense energy of the Big Bang is conveyed by the emission of radiation that we receive from all directions of the cosmos. This is **cosmic background radiation**, the vestiges of the conditions that ruled in the Universe following the Big Bang.

The **cosmic horizon** is the limit of the observable Universe; when we try to look farther, we hit a wall of dense, opaque gas. It will never be possible to observe the Big Bang directly.

THE ECHO OF THE BIG BANG

In 1989, the COBE satellite measured the background radiation and noted tiny variations in the temperature; these led to the formation of the large structures of the Universe, including galaxies.

Without telescopes to substitute for the naked eye, the most important astronomical discoveries would have been inconceivable. By enabling us to plumb the depths of space, giant telescopes and radio telescopes (which detect invisible forms of light) have profoundly changed our very vision of the Universe. Thanks to telescopes, thousands of stars and galaxies have been cataloged, new planets have been discovered, and a multitude of unique phenomena (including quasars and black holes) have been observed.

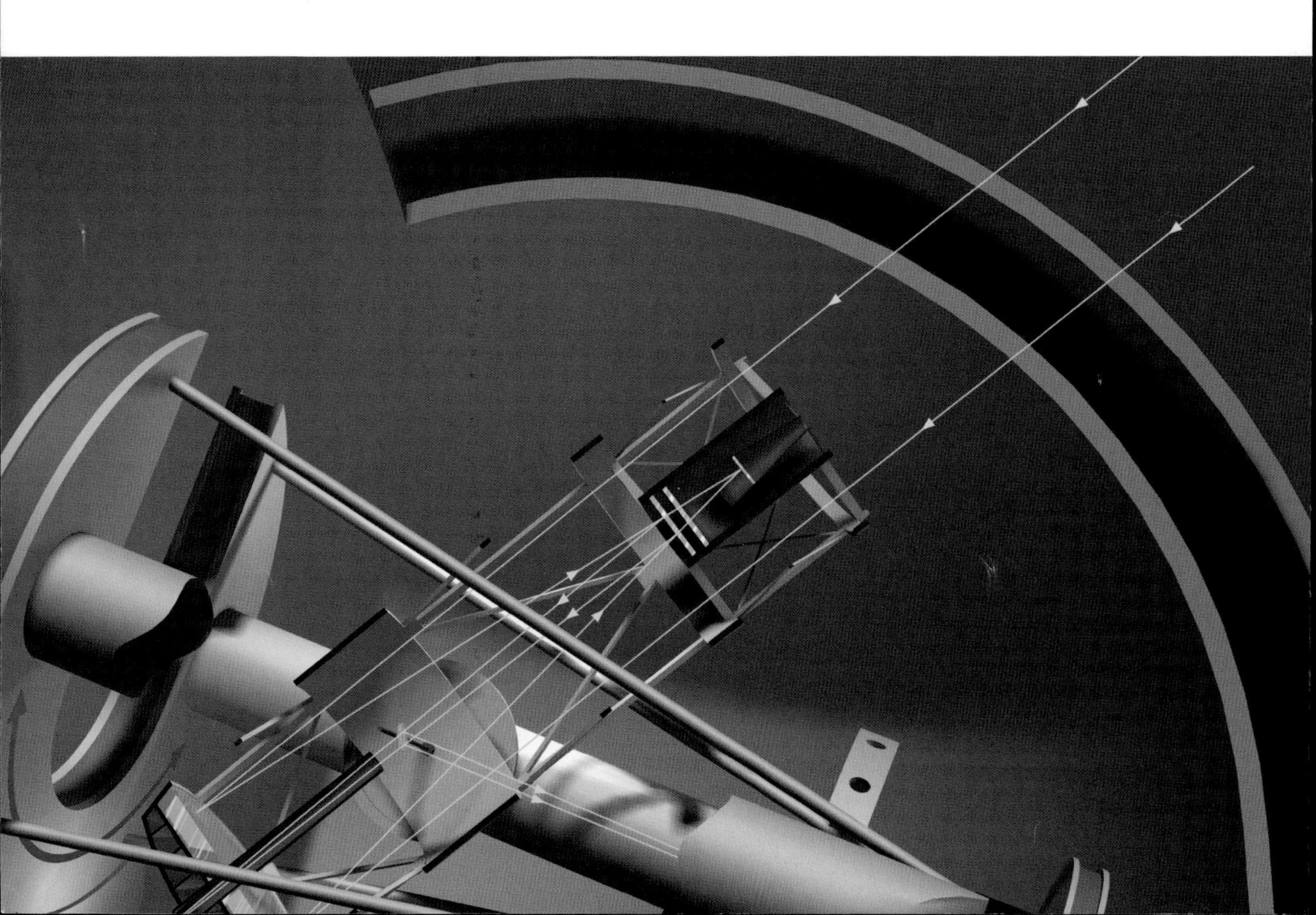

Astronomical Observation

The Electromagnetic Spectrum

When light is invisible

Our knowledge of the Universe does not come only from what we can see with the naked eye. Heavenly bodies emit energy that travels through space and arrives on Earth in the form of radiation of variable intensity, of which visible light is just a tiny component. Our eyes and conventional telescopes cannot see all of these invisible radiations, including radio waves, microwaves, infrared and ultraviolet radiation, X rays, and gamma rays, each of which has a different wavelength and frequency.

Recent advances in astronomy are due in large part to our expanding knowledge of these various forms of radiation. For example, in observing our own galaxy, we obtain different information and images depending on the types of waves we receive and analyze.

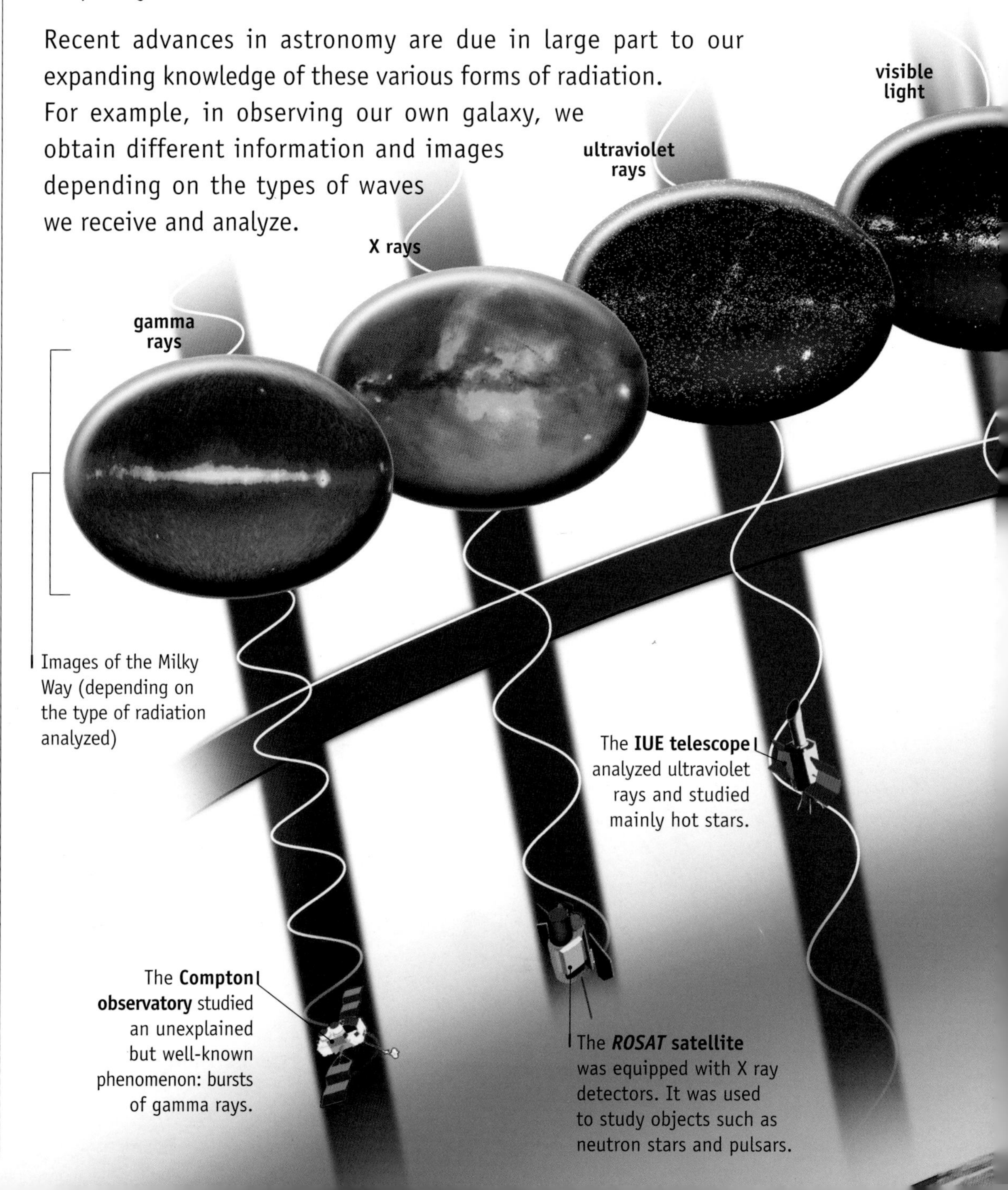

Images of the Milky Way (depending on the type of radiation analyzed)

The **IUE telescope** analyzed ultraviolet rays and studied mainly hot stars.

The **Compton observatory** studied an unexplained but well-known phenomenon: bursts of gamma rays.

The ***ROSAT*** **satellite** was equipped with X ray detectors. It was used to study objects such as neutron stars and pulsars.

OBSERVATORIES FOR DIFFERENT WAVE FORMS

Earth's atmosphere filters radiation from space, some of which is very high-energy and hazardous to all forms of life. Only visible light ❶ and radio waves ❷ reach the surface of our planet (along with some ultraviolet rays and infrared waves). We must therefore use various observatories placed in orbit to study the other kinds of radiation.

Telescopes

Concentrators of light

The invention of the telescope truly revolutionized our view of the Universe. For millennia, our ancestors studied the skies with the naked eye, without really seeing much. Between 1609 and 1612, using a small refracting telescope, Galileo discovered that the surface of the Moon is studded with craters and mountains, that there are spots on the surface of the Sun, and that the Milky Way is made of a huge number of stars!

Even today, experts observe the heavens using a reflecting telescope, a tube that gathers light coming from a heavenly body and concentrates it, using mirrors, into a given point.

TYPES OF TELESCOPES

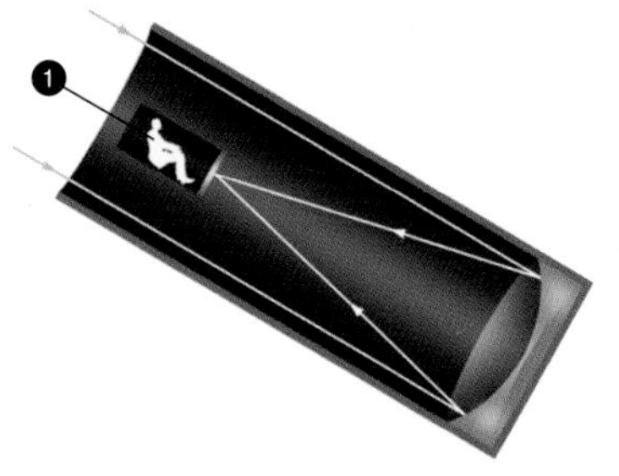

In a large **prime focal plane** telescope, the observer sits in an observer's cage ❶ installed in the tube and looks directly at objects in the prime focus.

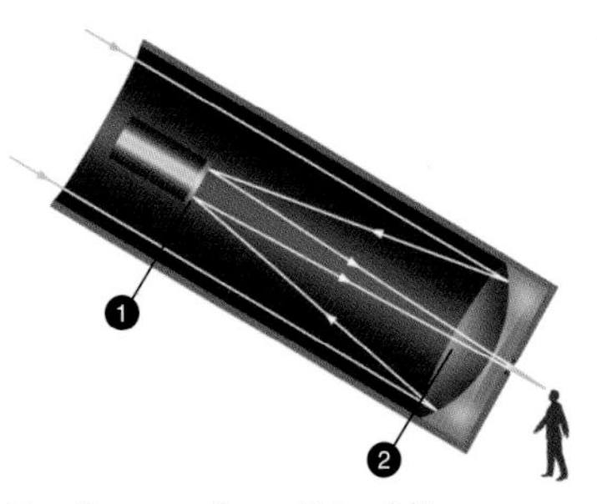

The **Cassegrain** or **Schmidt-Cassegrain** telescope uses a secondary mirror ❶, which reflects light back through a hole in the center of the primary mirror ❷.

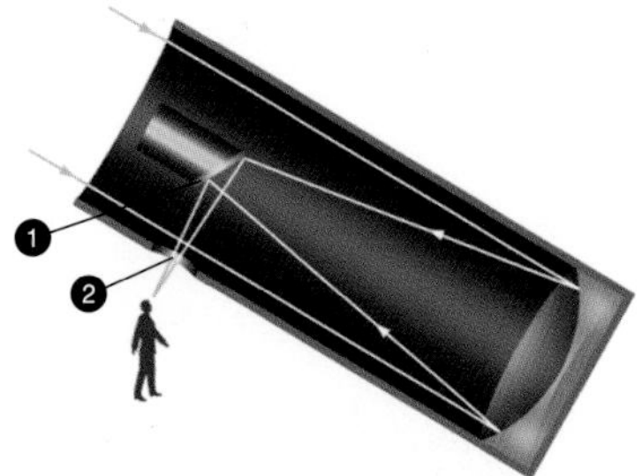

The **Newtonian** telescope deflects light with a secondary mirror ❶, inclined at 45°, toward an eyepiece ❷ on the side of the apparatus.

The **finder** is used to locate the objects to be observed.

The **eyepiece** is a magnifying lens that is used to look at the image formed at the prime focus.

A wide-aperture telescope captures more light and makes a clearer image than a narrow telescope, thus allowing heavenly bodies with lower luminosity to be seen.

REFLECTION

In a **reflecting telescope**, light ❶ is gathered by the objective, which is a concave primary mirror ❷ located at the bottom of the tube. It is concentrated into a focal point ❸ in front of the mirror (the prime focus). Then, the light is intercepted and reflected again, by a small flat mirror ❹, to the eyepiece ❺ on the side of the tube.

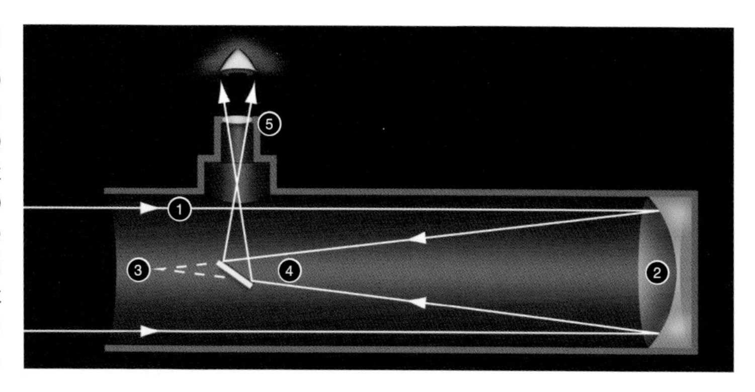

REFRACTION

In a **refracting telescope**, the light ❶ first passes through the objective ❷, then converges at the prime focus ❸. The image thus formed is channeled by a small mirror at a 45° angle ❹, which deflects the light to the eyepiece ❺.

The **declination adjustment** allows the telescope to be positioned vertically, relative to the equator.

The **right ascension adjustment** allows the telescope to track objects in a direction that is parallel to the equator.

Unlike a reflecting telescope, which uses mirrors, a **refracting telescope** concentrates light from space with lenses. More costly, and with crisper images, it is still used today by amateurs.

The First Astronomical Observatories

Seeing better and farther

In 1917, the greatest telescope designer of all time, George Hale, built a telescope with a mirror 8.2 feet (2.5 m) in diameter on top of Mount Wilson, in California. It was from this major observatory that Edwin Hubble made most of his discoveries about the immensity of the Universe. The Mount Palomar observatory followed in its footsteps in 1948, and a great number of this century's astronomical discoveries have been made there.

Located on top of mountains, where they are sheltered under huge domes that can open and pivot on their axis, these giant telescopes enable astronomers to study the skies as never before.

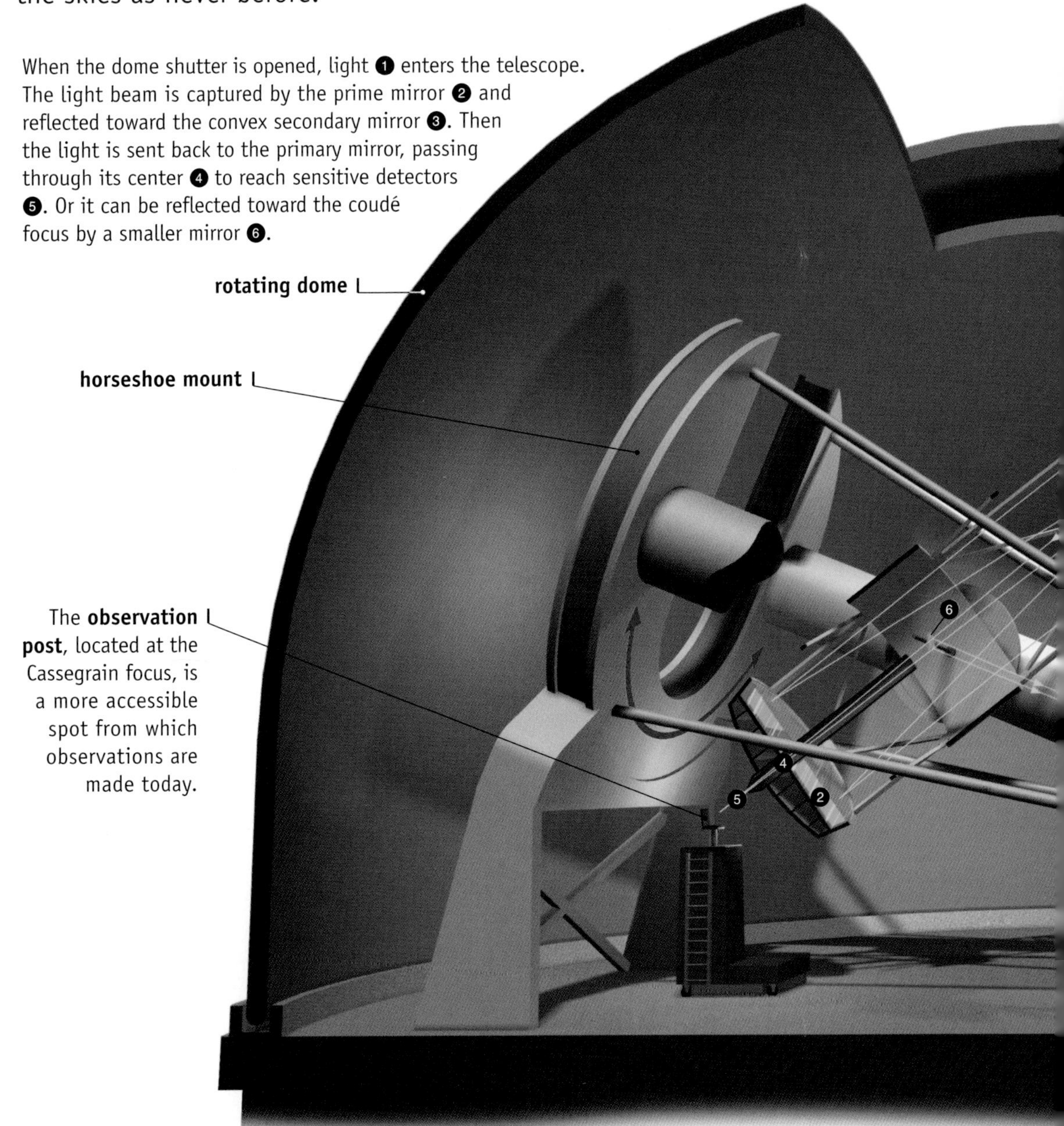

When the dome shutter is opened, light ❶ enters the telescope. The light beam is captured by the prime mirror ❷ and reflected toward the convex secondary mirror ❸. Then the light is sent back to the primary mirror, passing through its center ❹ to reach sensitive detectors ❺. Or it can be reflected toward the coudé focus by a smaller mirror ❻.

A REMARKABLE INNOVATION

Thanks to **charge-coupled devices (CCDs)**, electronic chips much more sensitive to light than a photographic plate, telescopes now capture extremely faint images with a shorter exposure time. The development of the CCD camera was a great leap in astronomical observation.

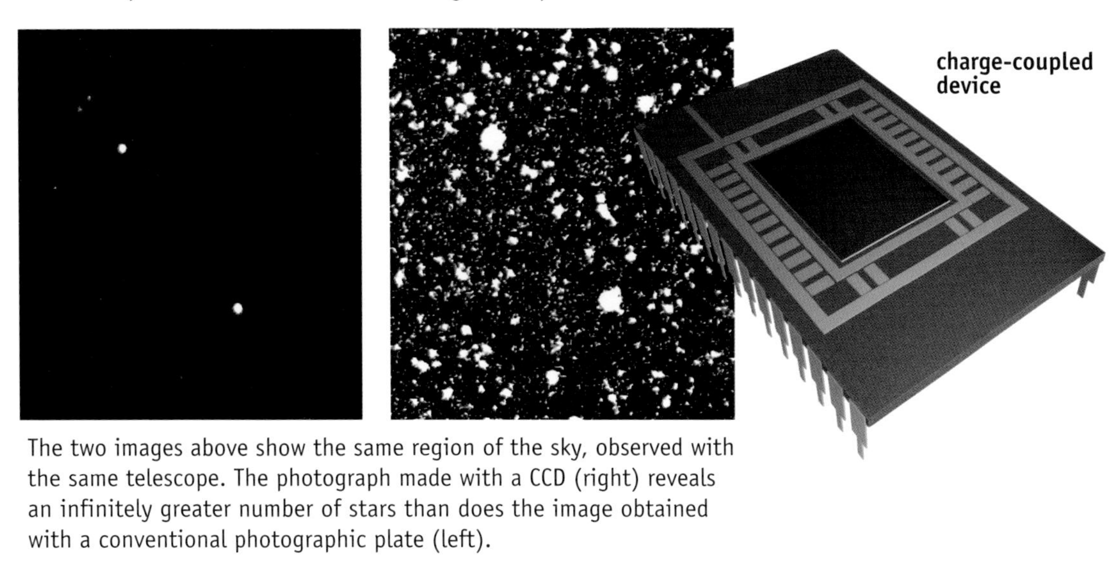

The two images above show the same region of the sky, observed with the same telescope. The photograph made with a CCD (right) reveals an infinitely greater number of stars than does the image obtained with a conventional photographic plate (left).

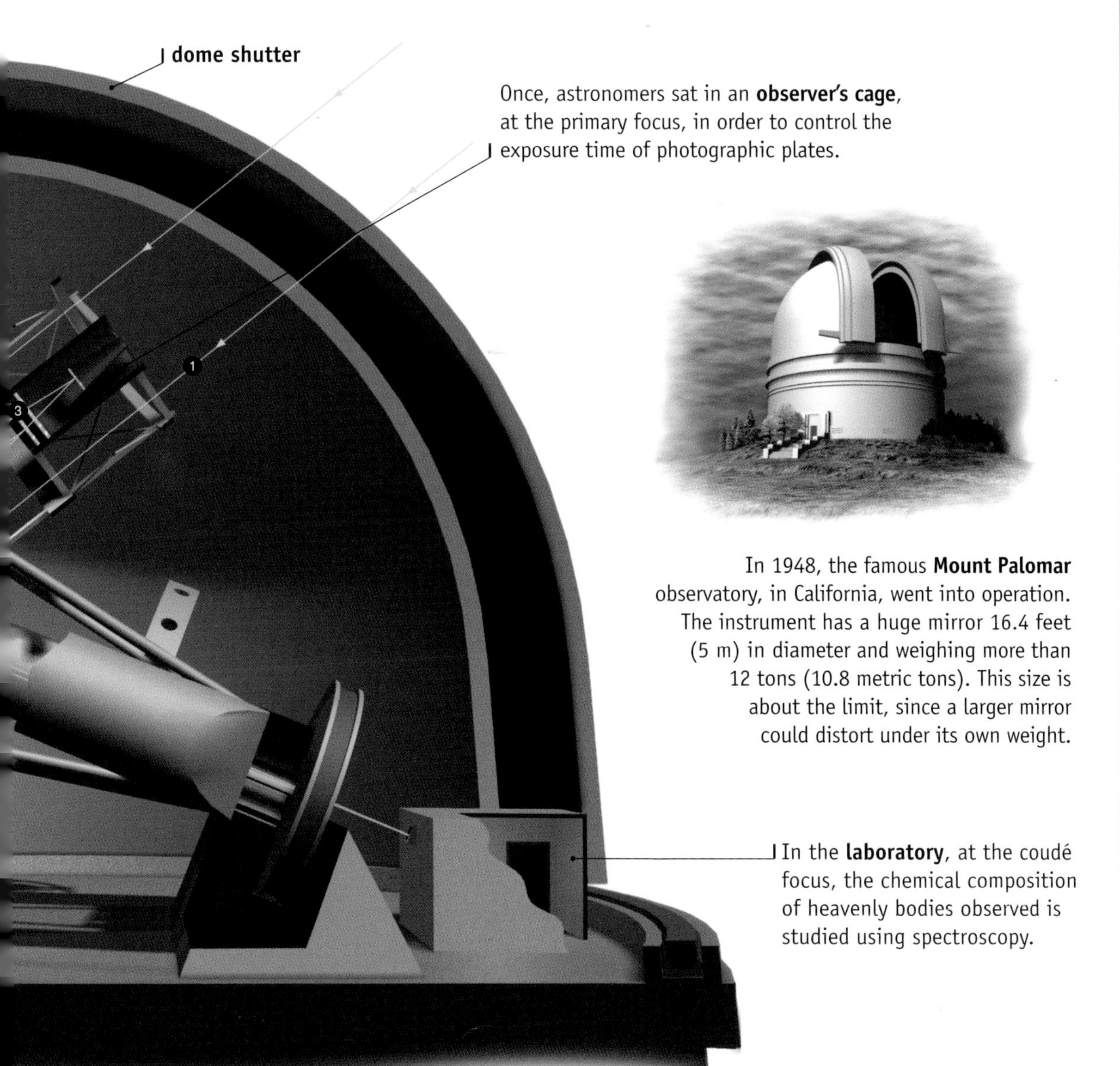

Once, astronomers sat in an **observer's cage**, at the primary focus, in order to control the exposure time of photographic plates.

In 1948, the famous **Mount Palomar** observatory, in California, went into operation. The instrument has a huge mirror 16.4 feet (5 m) in diameter and weighing more than 12 tons (10.8 metric tons). This size is about the limit, since a larger mirror could distort under its own weight.

In the **laboratory**, at the coudé focus, the chemical composition of heavenly bodies observed is studied using spectroscopy.

A New Generation of Telescopes

Observatories that are more and more powerful

In the 1970s, a whole new generation of telescopes was on the horizon. Armed with a number of very accurately coordinated mirrors, the telescopes function like one huge mirror. To avoid light pollution from big cities, these large observatories are set up on peaks of mountains in the middle of deserts, or on islands in the ocean. The first multiple-mirror telescope, on top of Mount Hopkins, Arizona, was inaugurated in 1979.

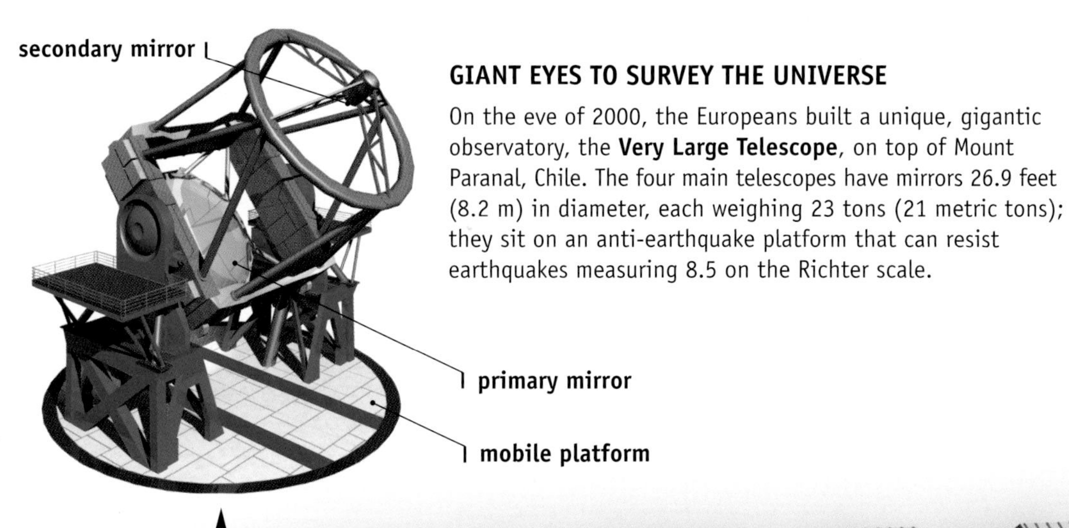

GIANT EYES TO SURVEY THE UNIVERSE

On the eve of 2000, the Europeans built a unique, gigantic observatory, the **Very Large Telescope**, on top of Mount Paranal, Chile. The four main telescopes have mirrors 26.9 feet (8.2 m) in diameter, each weighing 23 tons (21 metric tons); they sit on an anti-earthquake platform that can resist earthquakes measuring 8.5 on the Richter scale.

A GIANT ON TOP OF A MOUNTAIN

The huge **Keck** Telescope, in Hawaii, has thirty-six hexagonal mirrors 35 inches (90 cm) per side, forming a single reflector 32.8 feet (10 m) across. This telescope's resolution is four times better than that of the Mount Palomar telescope.

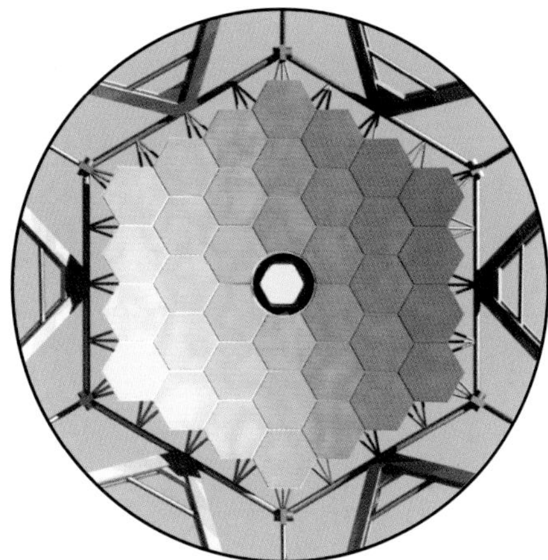

The light beams captured by each telescope are routed through an **underground tunnel**.

MULTIPLYING THE POWER OF TELESCOPES

A telescope's magnifying power is multiplied by the use of interferometry, a technique that increases the resolution of images. The light beams reflected by each telescope ❶ are directed by mirrors mounted on trolleys ❷ that move on tracks ❸ in an underground tunnel. The light beams are combined ❹ in the laboratory ❺ to obtain the precision of a mirror 394 feet (120 m) in diameter.

Three **auxiliary telescopes** 5.9 feet (1.8 m) in diameter can be moved to different positions to increase the accuracy of the observations.

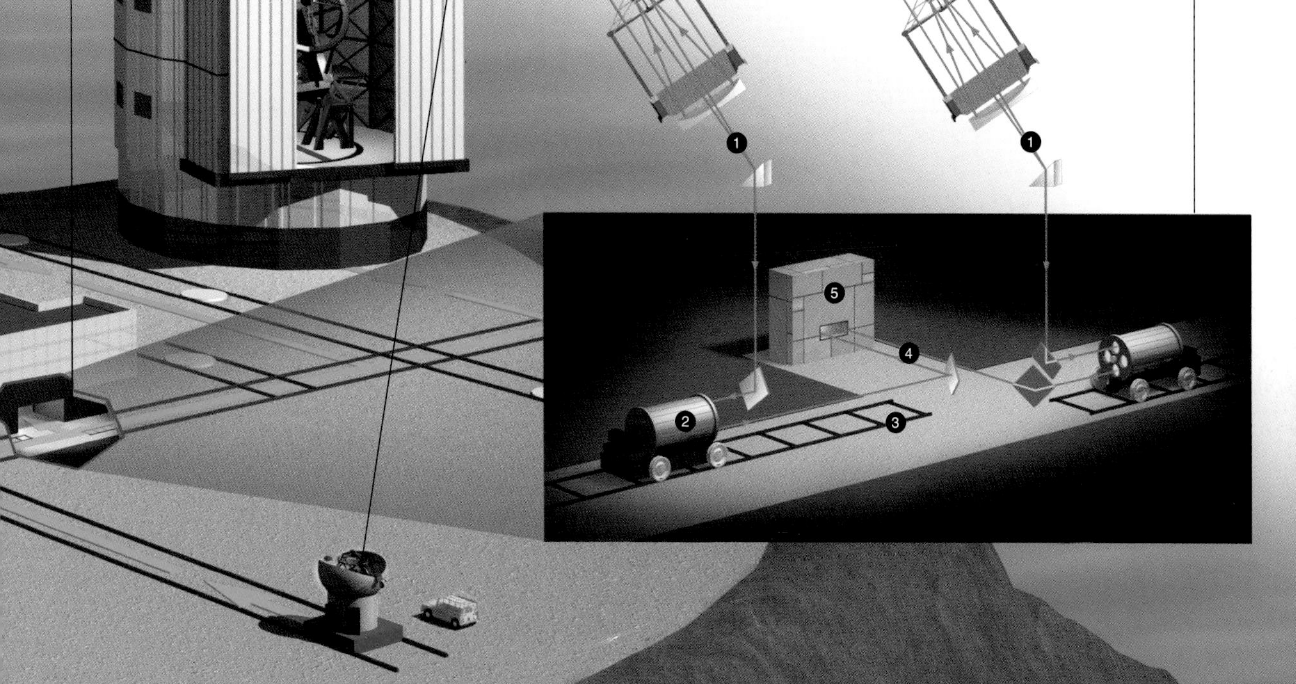

The Hubble Space Telescope

Beyond the clouds

The Hubble Space Telescope is one of the most important astronomical instruments of all time. The great advantage of this telescope is that it is above Earth's atmosphere, which filters and distorts the light from heavenly bodies. Placed in orbit in 1990, at 375 miles (600 km) altitude, the telescope transmits photographs of incomparable clarity and enables scientists to see farther than they could with any other astronomical instrument.

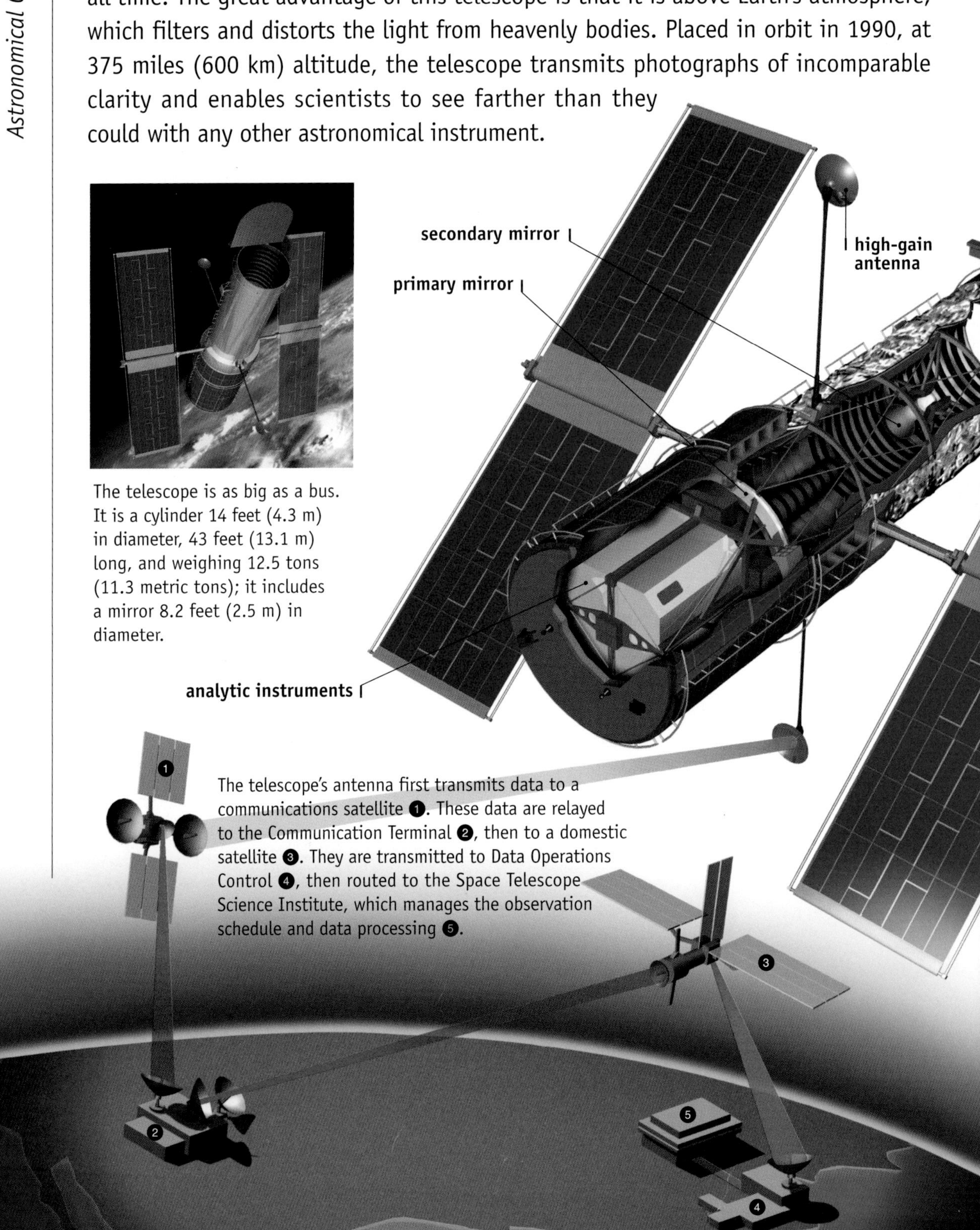

The telescope is as big as a bus. It is a cylinder 14 feet (4.3 m) in diameter, 43 feet (13.1 m) long, and weighing 12.5 tons (11.3 metric tons); it includes a mirror 8.2 feet (2.5 m) in diameter.

The telescope's antenna first transmits data to a communications satellite ❶. These data are relayed to the Communication Terminal ❷, then to a domestic satellite ❸. They are transmitted to Data Operations Control ❹, then routed to the Space Telescope Science Institute, which manages the observation schedule and data processing ❺.

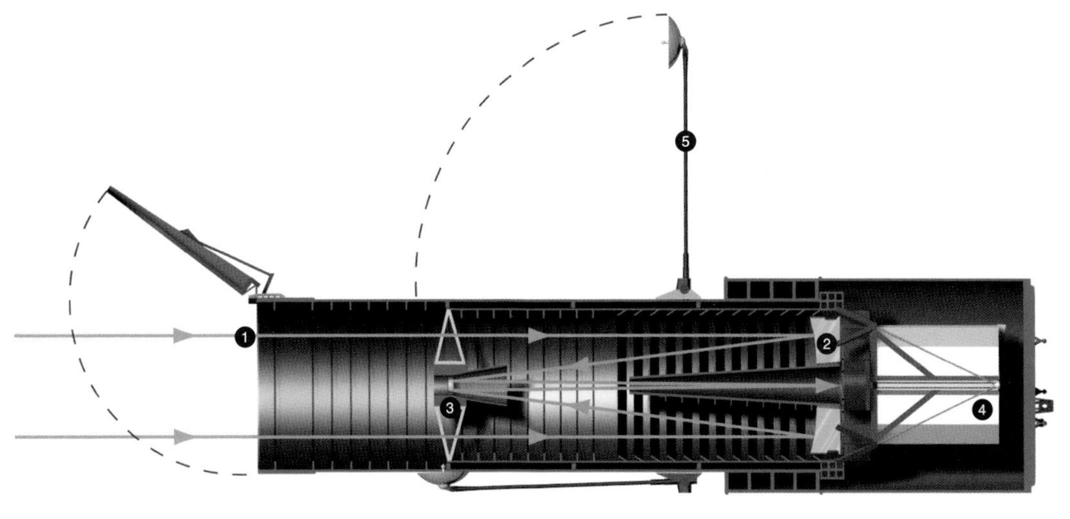

Light rays pass through the cylinder ❶ and are reflected by the primary mirror ❷ toward the secondary mirror ❸. This mirror sends the image to the analytic instruments ❹ (which include two cameras). The data are then retransmitted via an antenna ❺.

aperture door

REDISCOVERING THE UNIVERSE

Since it was launched, the Hubble Space Telescope has supplied us with more than 270,000 pictures of 13,600 heavenly bodies. These images have already had a profound impact on our concept of the Universe; as it probes the depths of space, the telescope has shown us how different the Universe was when it was young. Its ultimate mission, which is to determine the scope, size, and age of the Universe, could have plenty of surprises in store before the Hubble Space Telescope is retired in 2010.

Solar panels provide electricity for the telescope.

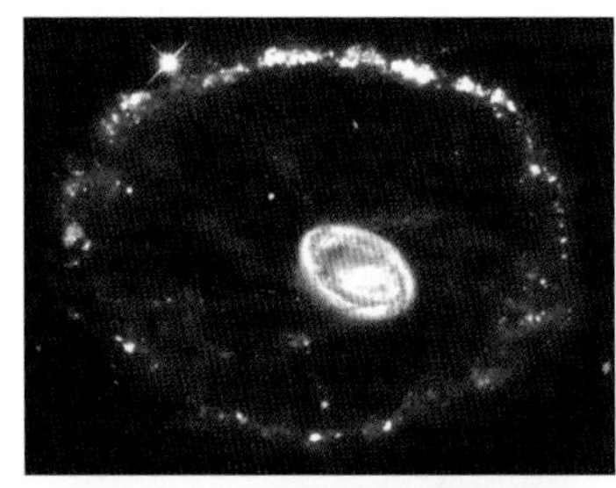

The Hubble telescope has revealed many unprecedented phenomena, including large-scale star formation after a huge shock wave originating in the heart of the **Cartwheel Galaxy** (400 million light-years away).

Hubble showed us the **Eagle Nebula**, which is among the most spectacular views in the Universe. At the peak of huge columns of dust, several light-years long, stars are being born.

The extremely powerful Hubble telescope photographed a **tiny piece of the Universe** (to our eyes the size of a coin 82 feet [25 m] away) in which more than fifteen hundred galaxies of all shapes and ages were counted.

Eta Carinae, one of the most massive and unstable of stars, is surrounded by an incandescent envelope created by the constant ejection of matter, as this photograph taken by Hubble shows.

Radio Telescopes

A new window on the Universe

The Universe is chock-full of objects that cannot be observed with an optical telescope, even a very powerful one. In fact, visible light is not the only form of light that comes from space. Heavenly bodies also emit radio light; some actually emit more radio waves than light waves. Radio telescopes work on the same principle as optical telescopes, but they are designed specifically to capture and concentrate invisible radio waves. These instruments can be used both day and night, and they can "see" through clouds.

"VIEWING" THE INVISIBLE

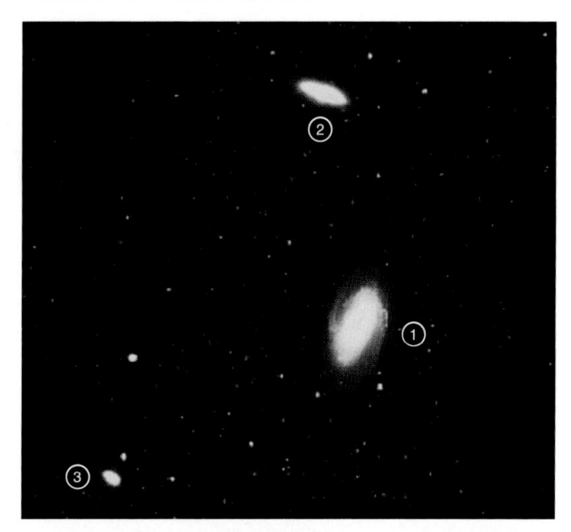

A photograph in **visible light** seems to show that the impressive galaxy M81 ❶, the galaxy M82 ❷, and the small, irregularly shaped NGC 3077 ❸ are three independent heavenly bodies.

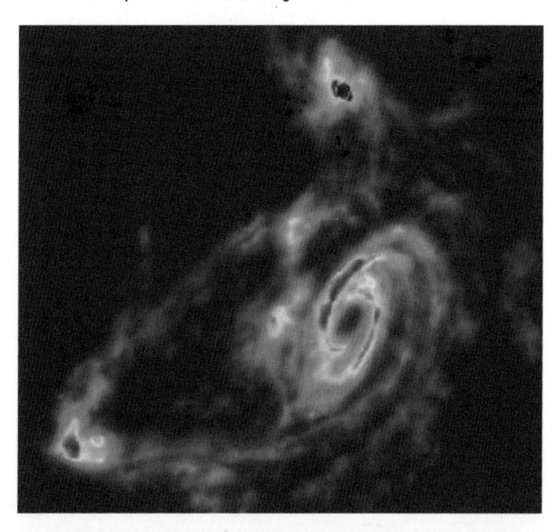

A photograph in **radio light** shows that an immense hydrogen cloud links the three galaxies.

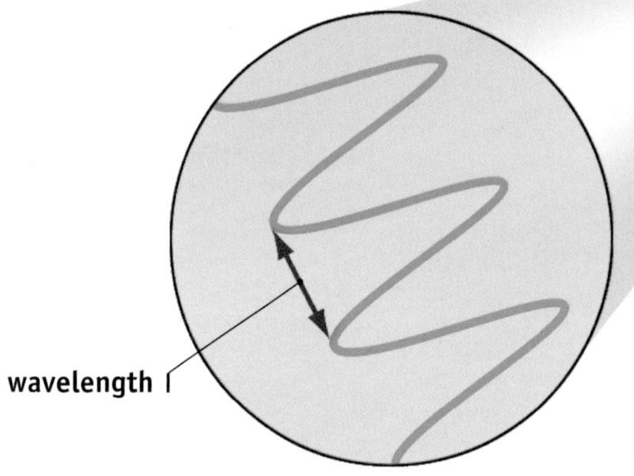

Radio telescopes capture **radio waves**, whose length varies from fractions of an inch (several millimeters) to 66 feet (20 m).

THE LARGEST RADIO TELESCOPE IN THE WORLD

Some stationary radio telescopes, in the form of giant parabolic surfaces, are built in valleys. The largest one is the famous **Arecibo Radio Telescope**, in Puerto Rico, which measures 1,000 feet (305 m) in diameter.

THE EFFELSBERG OBSERVATORY

Because radio waves are much longer than light waves, radio reflectors are usually very large. One of the largest **steerable parabolic reflectors** (the most common type of radio telescope) is located in Effelsberg, Germany; it is 328 feet (100 m) in diameter — the length of a football field.

Radio telescopes are large parabolic antennas that capture radio waves ❶ using a primary reflector ❷ that concentrates them toward a primary focus ❸ at the top of the antenna. The radio waves are then amplified by receivers ❹, or they can be sent to a secondary focus ❺, where they are amplified again ❻ and recorded and analyzed in a laboratory ❼.

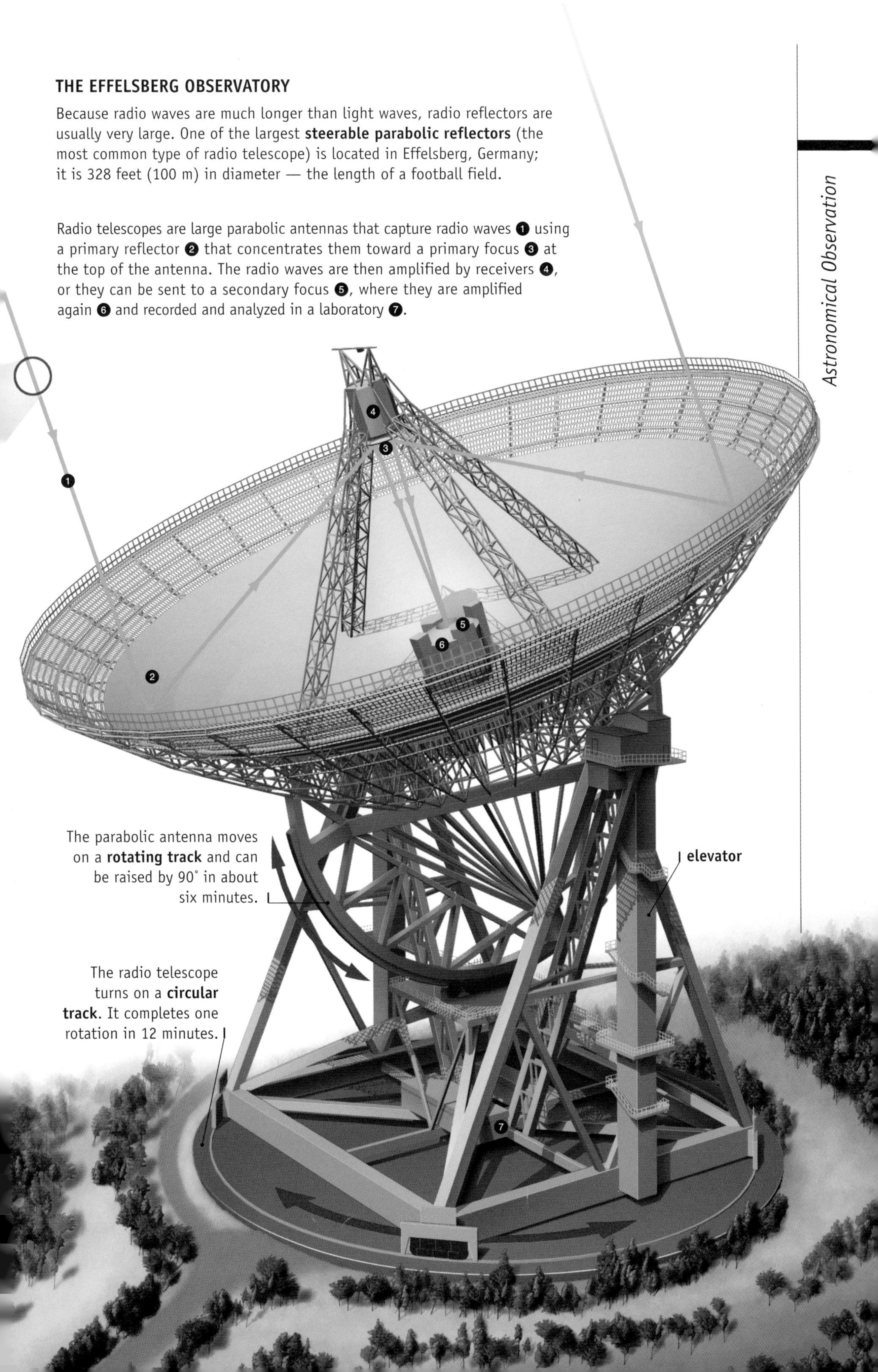

Life Elsewhere in the Universe

Are we alone?

We have been wondering if we are alone in the Universe for a very long time. Since the second half of the twentieth century, however, this speculation has become a science in itself. *Exobiology* is the discipline of determining the conditions necessary for life and the places where it might develop, using the technical means that might permit us to locate it.

DRAKE'S EQUATION

In 1961, a U.S. radioastronomer, Frank Drake, came up with an equation that theoretically estimates the probability of the existence of intelligent life in our galaxy. The formula he devised became the basis for discussion on the subject. Its aim is to calculate the number of communicating civilizations that might exist in the Milky Way and from which we might reasonably hope to receive a signal. The equation is as follows: $N = (R^*) \times (F_p) \times (N_e) \times (F_l) \times (F_i) \times (F_t) \times (L)$.

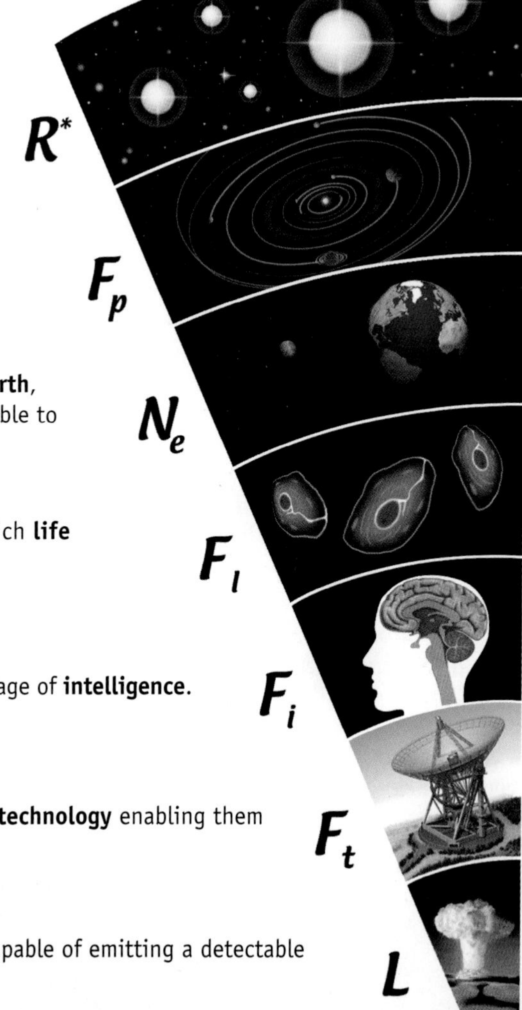

R^* is the formation rate of **stars** around which a civilization might develop. This number is a fraction of all the stars in the galaxy and excludes large stars whose life span is too short to enable evolution of a communicating civilization.

F_p is the fraction of these stars that have a **planetary system**.

N_e corresponds to the number of planets resembling **Earth**, located in a habitable zone featuring conditions favorable to the creation of life.

F_l is equivalent to the fraction of these planets on which **life** could develop.

F_i is the fraction of planets where life has reached the stage of **intelligence**.

F_t is the fraction of civilizations that have developed a **technology** enabling them to send signals in space.

Finally, L corresponds to the **lifetime** of civilizations capable of emitting a detectable radio signal.

N corresponds to the number of **communicating civilizations** in the Milky Way that could emit radio signals we could detect. This number varies greatly according to the values used for each of the above parameters; the number goes from 1 (our civilization) to millions, or even billions.

CONDITIONS NECESSARY FOR LIFE

For life to come into existence and develop, it must benefit from conditions similar to those on Earth. Basic materials, such as carbon and liquid water, must be found on the surface of a planet with an atmosphere, and a fairly stable environment must exist over hundreds of millions of years. Such a planet must also be neither too near to nor too far from a star that is burning slowly enough to give time for life to organize.

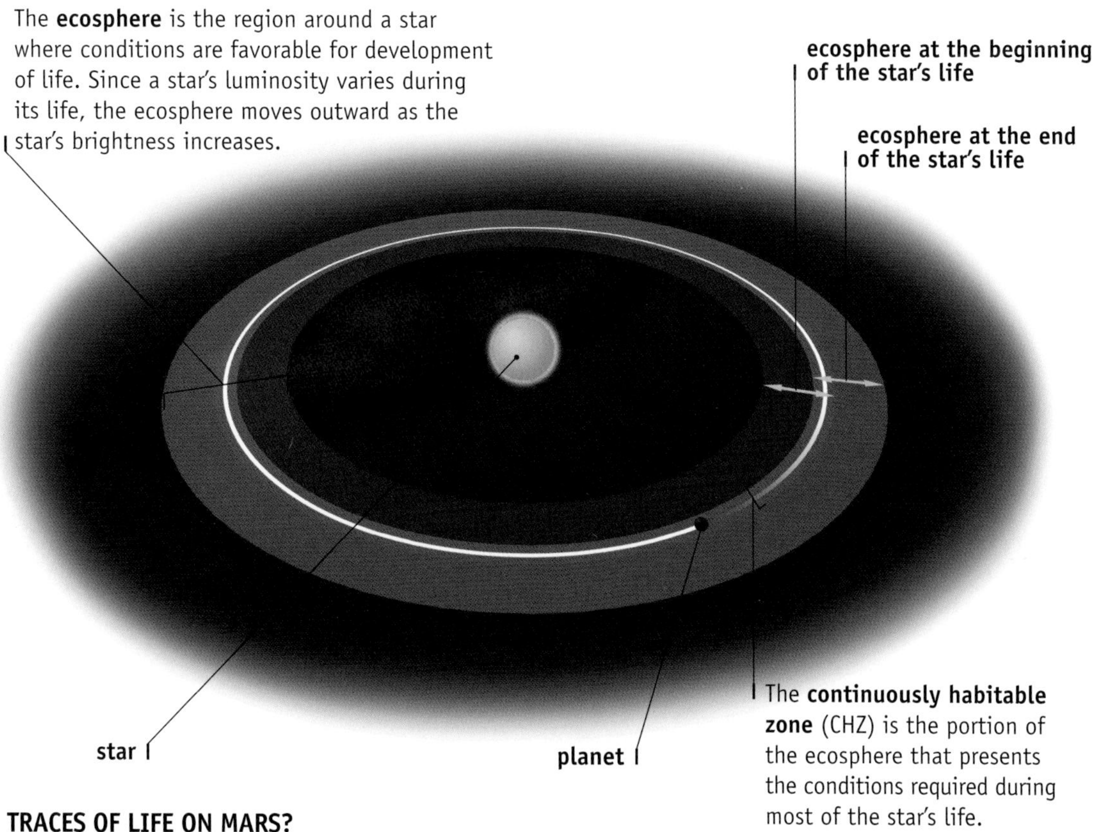

TRACES OF LIFE ON MARS?

In 1996, what may be traces of microfossils were discovered in a meteorite that came from Mars. It contained longiform structures just a few micrometers in length that resembled terrestrial bacteria.

Discovered in Antarctica, where it had fallen 13,000 years ago, the meteorite had crystallized on Mars 4.5 billion years earlier, when the planet was formed. The microfossils would have become lodged in it at a time when Mars was a warm, humid planet.

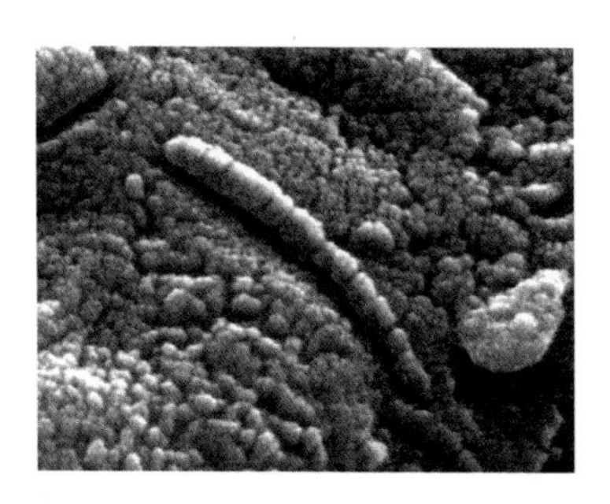

On Mars, **dried riverbeds** are vestiges of the time when the planet had a more temperate climate and could support life.

A MESSAGE TO EXTRATERRESTRIALS

In 1974, the Arecibo radio telescope sent a message in binary code toward a globular cluster in the constellation Hercules. The message will arrive at this cluster, which contains hundreds of thousands of stars, in 25,000 years.

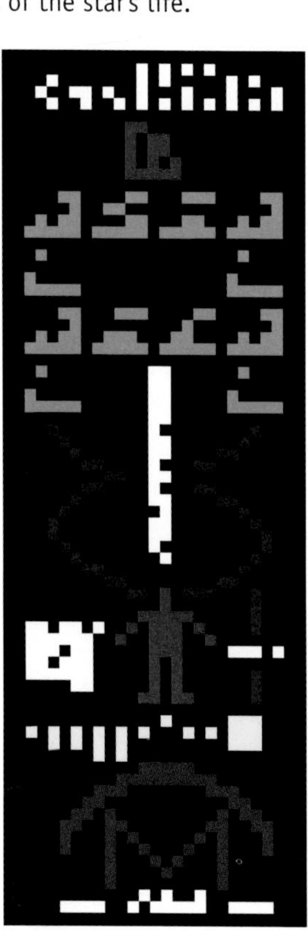

Discovering Extrasolar Planets

A little lesson in planetary humility

For a long time, it was thought that no planets existed outside of our Solar System. But for about fifty years, astronomers have been looking in the vicinity of nearby stars for extrasolar planets, called exoplanets. The first clue that exoplanets exist was discovered in 1984, when the *IRAS* satellite observed dust rings around some twenty stars. Such structures, which are no doubt similar to what our Solar System looked like when it was very young, indicate that planet formation is a much more frequent phenomenon than was previously believed.

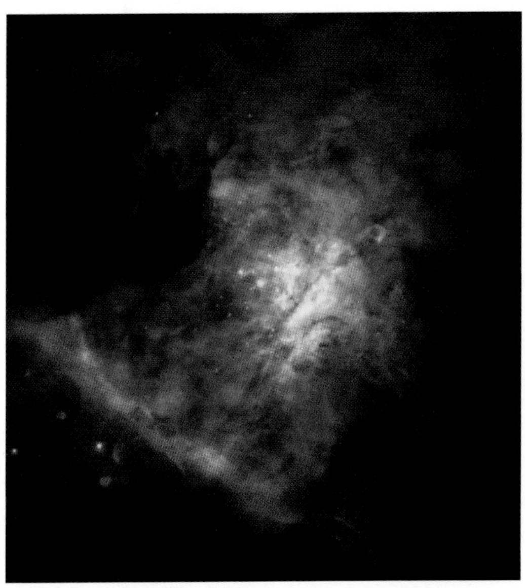

The **Orion Nebula** contains a number of young stars around which new planets might form.

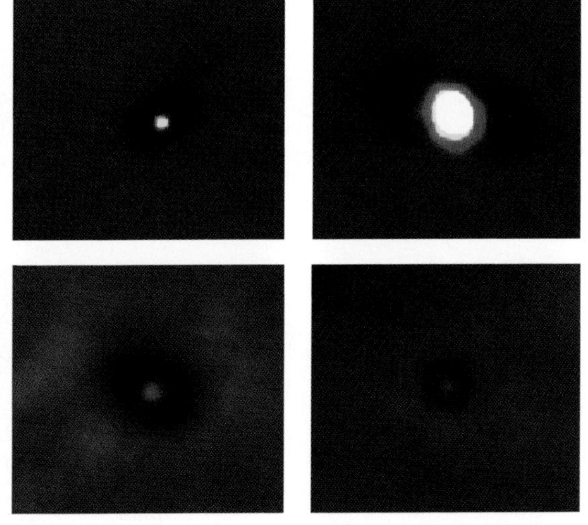

The Hubble Space Telescope observed disks made of gas and dust around more than 150 stars in the Orion Nebula. These are **protoplanetary disks**, which are probably planetary systems in the process of being formed.

A PLANET IN THE MAKING?

An object that could be a **protoplanet** and its star have been observed 450 light-years from Earth, in the constellation Taurus. This protoplanet (TMR-1C) may be two to three times the mass of Jupiter, the largest planet in our Solar System.

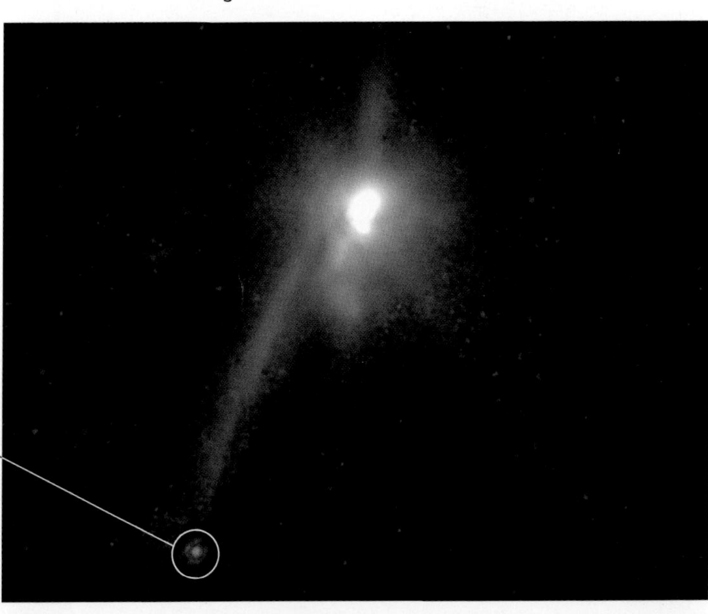

NEW PLANETS

Since 1995, the scrutiny of hundreds of stars has revealed the existence of planets revolving around stars comparable to our Sun. Most are located closer to their star than Earth is to the Sun, and they have a mass equal to or greater than Jupiter's (M Jup). They take anywhere from several days to several years to revolve around their stars.

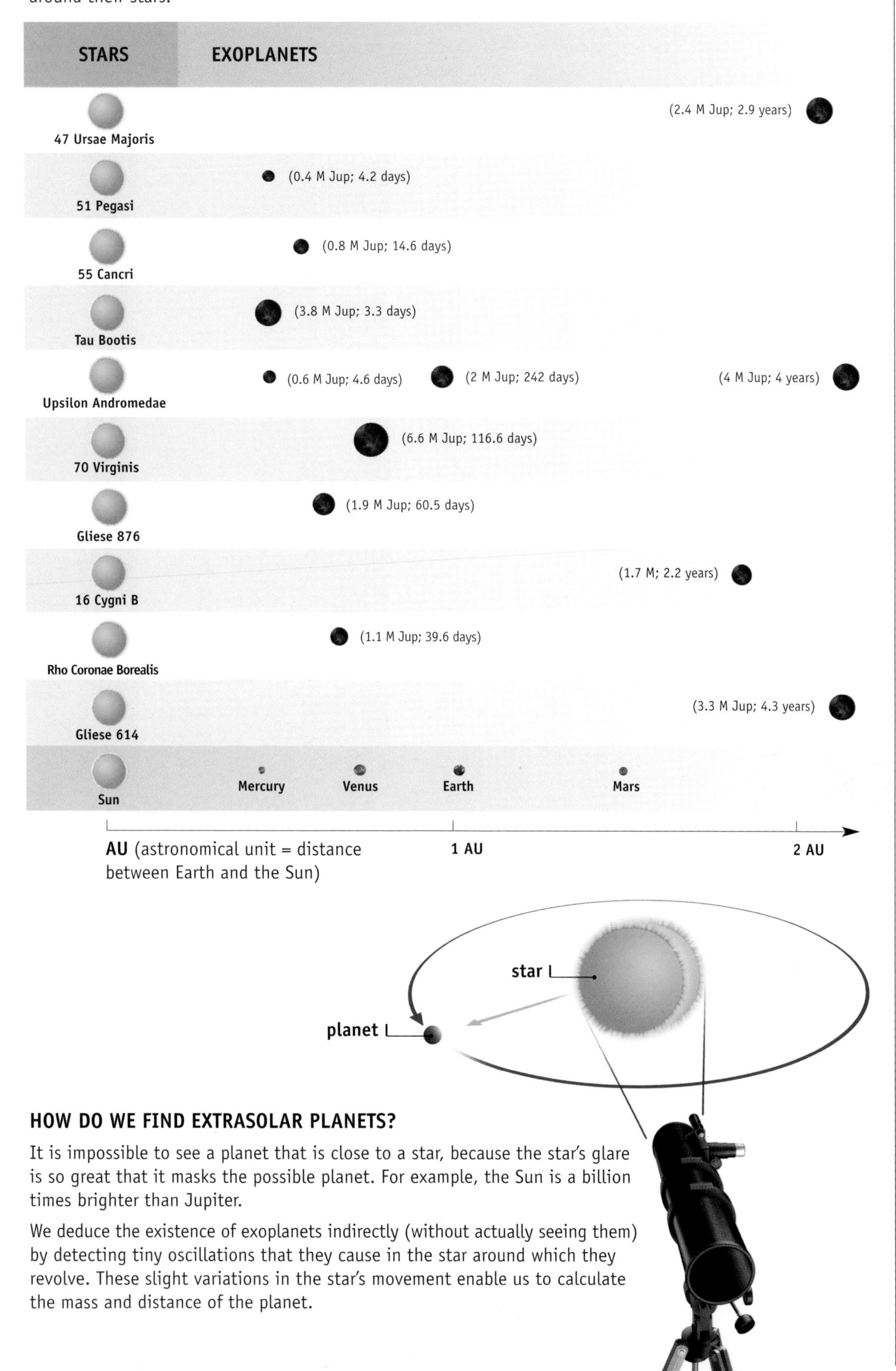

HOW DO WE FIND EXTRASOLAR PLANETS?

It is impossible to see a planet that is close to a star, because the star's glare is so great that it masks the possible planet. For example, the Sun is a billion times brighter than Jupiter.

We deduce the existence of exoplanets indirectly (without actually seeing them) by detecting tiny oscillations that they cause in the star around which they revolve. These slight variations in the star's movement enable us to calculate the mass and distance of the planet.

Similar to the giant eyes of telescopes, space probes are a marvelous way of increasing our knowledge of the cosmos. As they fly over hostile environments, sending back pictures, landing where human beings cannot go, they bring back data and samples for analysis. These fabulous devices add to our knowledge of planets, comets, asteroids, and many other bodies. Above all, they help us measure, to the extent that it is possible, the immeasurable space in our small corner of the Universe.

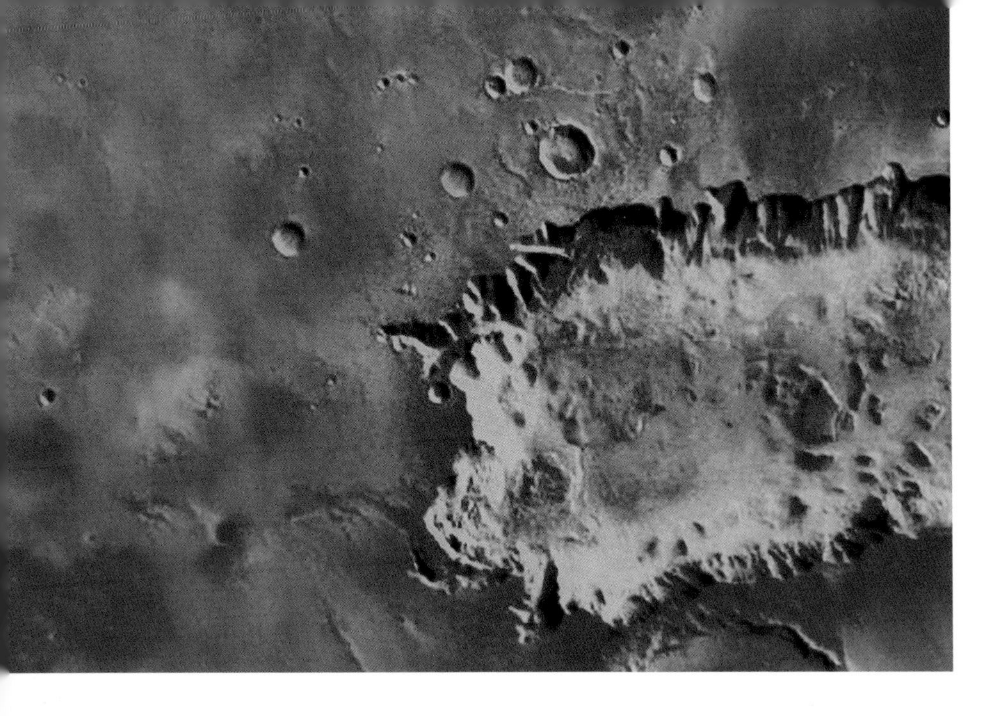

Space Exploration

Space Probes

The great explorers of modern times

They have names like *Pioneer, Voyager, Galileo, Magellan,* and *Ulysses*. They are the explorers of our times, successors to Marco Polo, Christopher Columbus, and Ferdinand Magellan, who explored our globe before the Renaissance. These modern explorers are robots; as substitutes for our eyes and other senses, they have transformed our ideas about the Solar System in a single generation.

THE THREE STAGES OF PLANETARY EXPLORATION

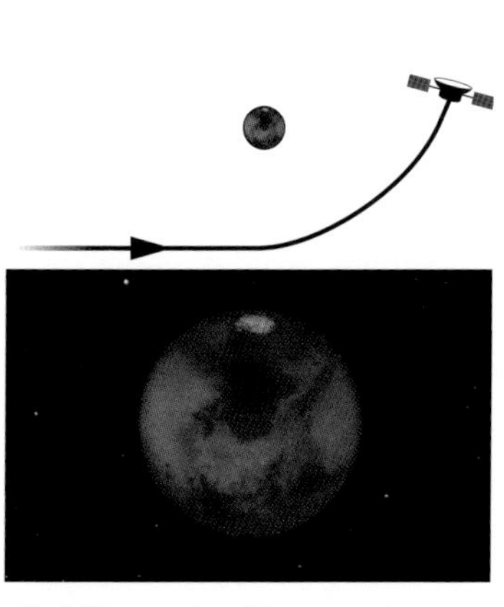

❶ A first probe flies past the planet, giving us a brief but spectacular view of it.

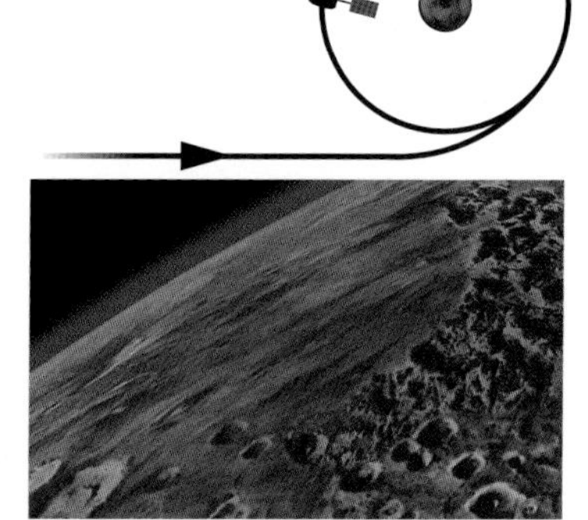

❷ Another probe is then placed in orbit around the planet for a more thorough examination, giving us a good overall view.

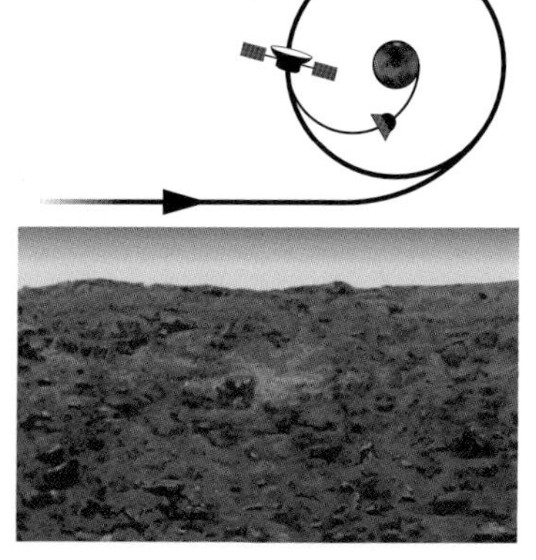

❸ Finally, a robot touches down on the planet's surface and provides a very detailed local view. This procedure has been followed for the Moon, Mars, and Venus.

THE EVOLUTION OF SPACE EXPLORATION

Advances in planetary exploration have been made in large part by technological progress. Radioisotope thermoelectric generators now provide the electricity for probes through nuclear reactions.

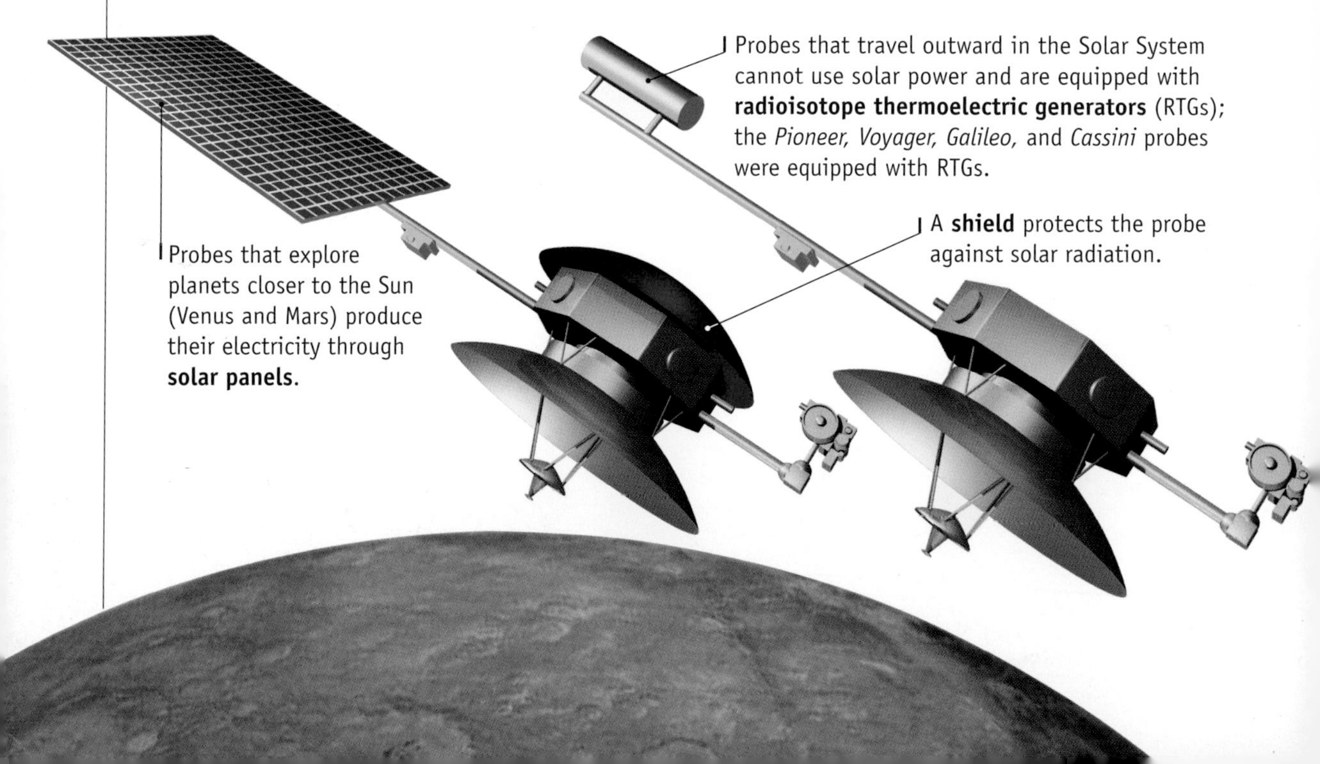

Probes that travel outward in the Solar System cannot use solar power and are equipped with **radioisotope thermoelectric generators** (RTGs); the *Pioneer, Voyager, Galileo,* and *Cassini* probes were equipped with RTGs.

A **shield** protects the probe against solar radiation.

Probes that explore planets closer to the Sun (Venus and Mars) produce their electricity through **solar panels**.

A TYPICAL SPACE PROBE

Space probes are among our most ingenious technological feats. Not only must they travel hundreds of millions of miles (hundreds of millions of km) in the hostile environment of interplanetary space, but they must also accomplish all mission maneuvers — including landing — using onboard computers, with no assistance from Earth. A typical probe has two modules: an orbiter and a lander.

THE ORBITER

After flying to a planet, the probe goes into orbit around it and examines it in detail for several months.

transmission antenna

A **compass** uses a guide star (such as Canopus) as a reference point to orient the probe.

The **camera** takes thousands of pictures of the planet, providing an overall view.

The **infrared thermal mapper** uses infrared rays to study the planet's surface and the composition of its atmosphere.

THE LANDER

The lander is designed to touch down on a planet's surface and study it. In a miniaturized structure, it combines power generators, chemical-analysis laboratories, television cameras, a meteorology station, and a computer center; on Earth, these would take up several floors of a building.

The directional **antenna** is always pointed toward Earth, to transmit scientific and photographic data.

Meteorological sensors measure temperature, atmospheric pressure, and the speed and direction of winds.

The **automated laboratory** analyzes the samples collected to determine their composition and check for any signs of life.

A **shovel** attached to a sampler boom collects soil samples to deposit in the automated laboratory.

Cameras take pictures of the planet's surface.

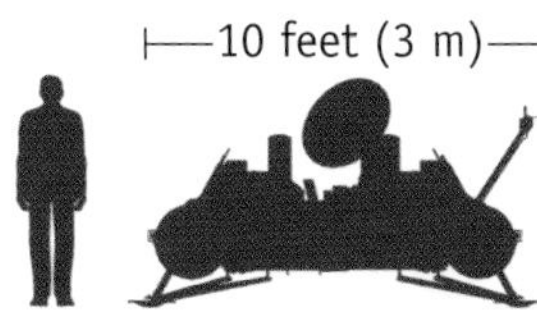

Pioneer 10 and 11

The first great voyagers

Pioneer 10 and *Pioneer 11* were the first probes to venture beyond the orbit of Mars. These small robots (570 pounds [260 kg]) were launched in March 1972 and April 1973. *Pioneer 10* was the first probe to enter the asteroid belt, which it crossed uneventfully. In December 1973 it passed within 80,700 miles (130,000 km) of Jupiter and transmitted the first close-up images of the giant planet; it observed Jupiter's intense magnetic fields and discovered that the planet has no solid surface.

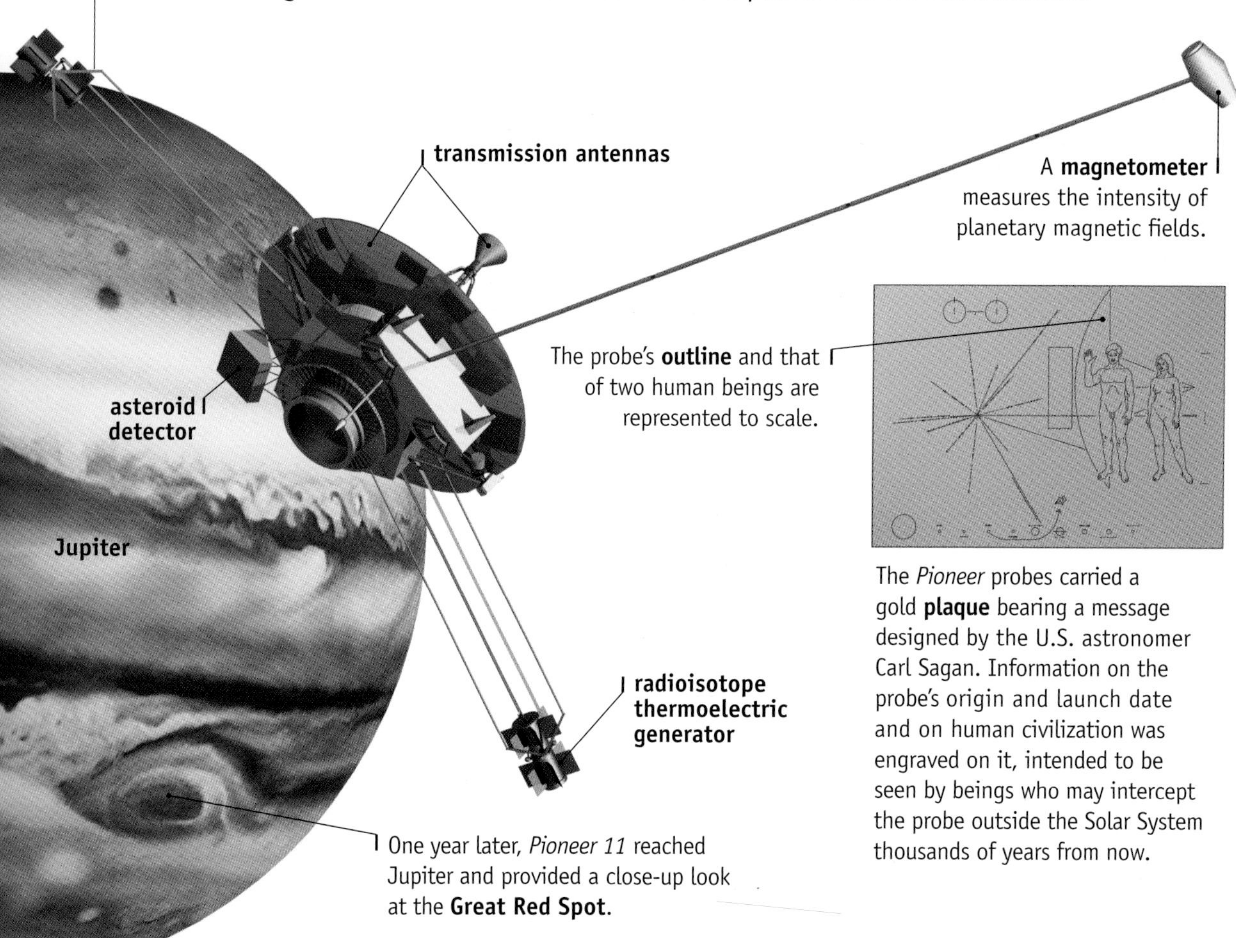

A **magnetometer** measures the intensity of planetary magnetic fields.

The probe's **outline** and that of two human beings are represented to scale.

The *Pioneer* probes carried a gold **plaque** bearing a message designed by the U.S. astronomer Carl Sagan. Information on the probe's origin and launch date and on human civilization was engraved on it, intended to be seen by beings who may intercept the probe outside the Solar System thousands of years from now.

One year later, *Pioneer 11* reached Jupiter and provided a close-up look at the **Great Red Spot.**

LEAVING THE SOLAR SYSTEM

After flying over Jupiter, the *Pioneer* probes continued on their way, exploring the outer reaches of the Solar System. *Pioneer 10* completed its scientific mission in March 1997. It is now traveling toward the star Aldebaran (sixty-eight light-years away), which it should reach in 2 million years. *Pioneer 11* stopped transmitting in November 1995 and is on its way toward Aquila, where it may pass near a star in 4 million years.

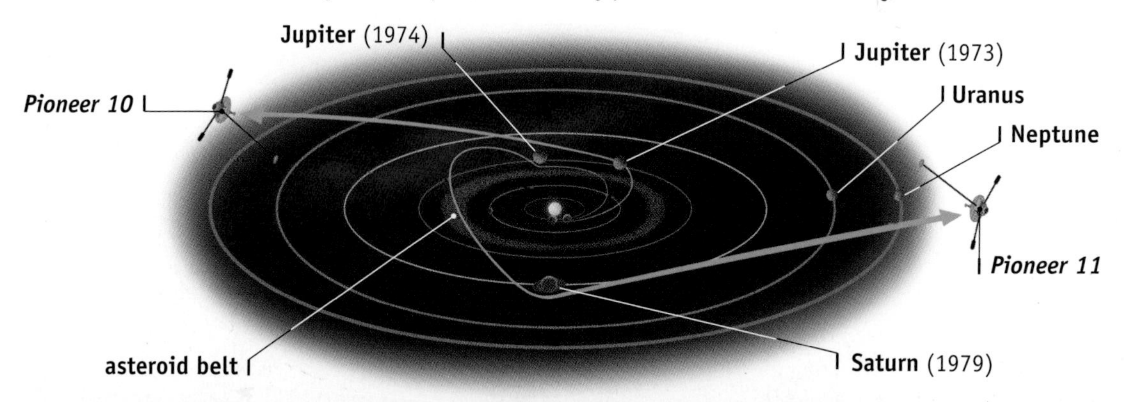

Viking

Discovering the fascinating deserts of Mars

After the Moon, Mars was the next target for space probes; over the course of thirty years, three dozen probes were sent to the Red Planet. In 1965, *Mariner 4* revealed that Mars resembles a barren desert. Then, in November 1971, *Mariner 9*, the first probe to be put in orbit, found areas shaped by water and many incredible landscapes, including the giant volcano Olympus Mons and the extraordinary canyon Valles Marineris.

In 1975, NASA launched the *Viking 1* and *Viking 2* probes, each consisting of an orbiter, which observed the planet from orbit, and a lander, which touched down on the surface. The landers did not move, however; *Sojourner* would accomplish this feat twenty years later.

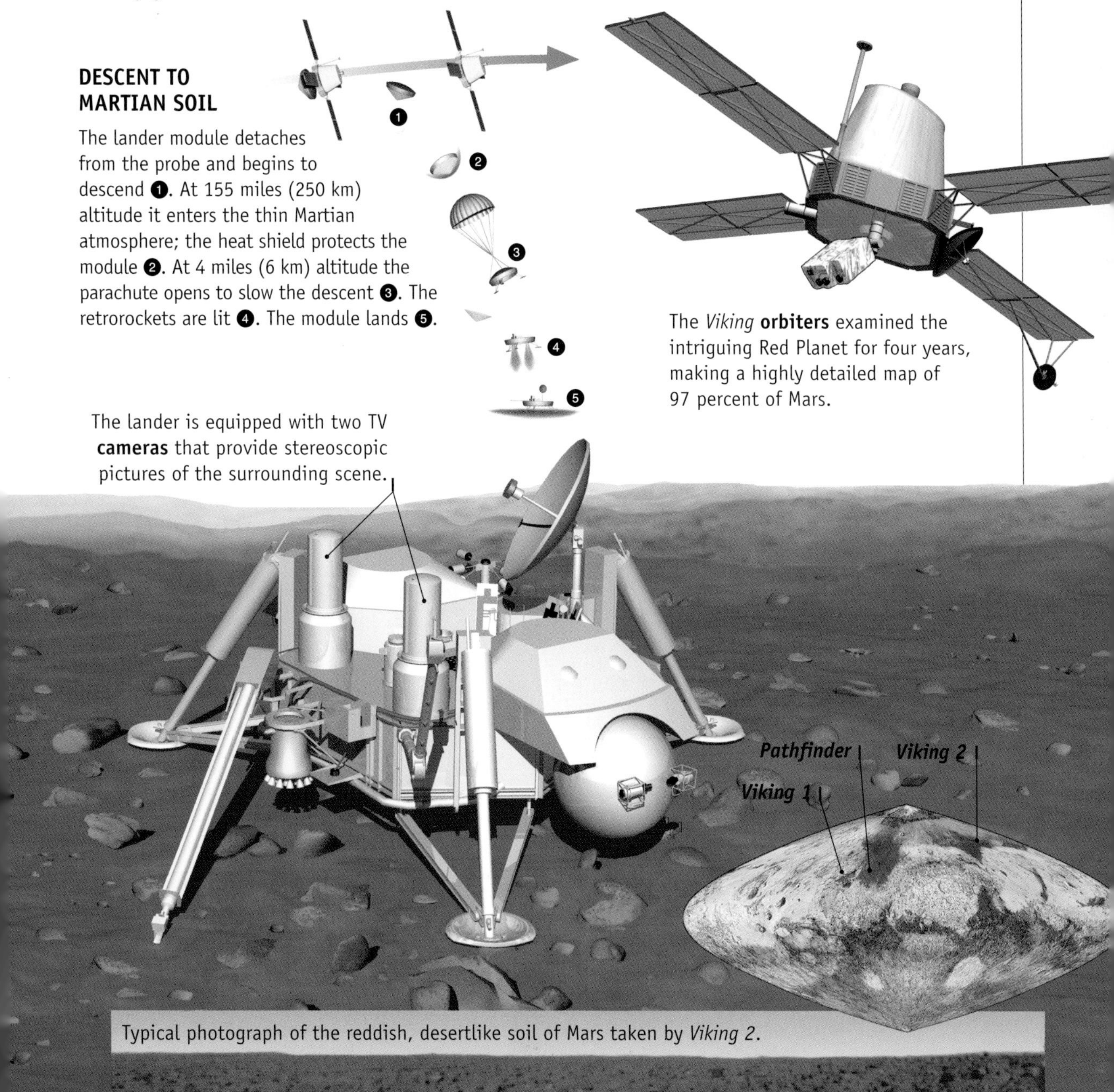

DESCENT TO MARTIAN SOIL

The lander module detaches from the probe and begins to descend ❶. At 155 miles (250 km) altitude it enters the thin Martian atmosphere; the heat shield protects the module ❷. At 4 miles (6 km) altitude the parachute opens to slow the descent ❸. The retrorockets are lit ❹. The module lands ❺.

The *Viking* **orbiters** examined the intriguing Red Planet for four years, making a highly detailed map of 97 percent of Mars.

The lander is equipped with two TV **cameras** that provide stereoscopic pictures of the surrounding scene.

Typical photograph of the reddish, desertlike soil of Mars taken by *Viking 2*.

Voyager

Grand tour of the Solar System

In the late 1970s a particular alignment of the giant planets (which takes place every 175 years) started U.S. scientists thinking about an ambitious project called the Grand Tour. Because it was too expensive, the project was abandoned, but in 1977 NASA launched two *Voyager* probes whose objective was to fly over Jupiter and Saturn.

The *Voyager* missions are now over, in principle. In February 1999, *Voyager 1* was 6.8 billion miles (10.9 billion km) from Earth, making it the most distant artificial object in space. It is expected that both probes will continue to function until 2020.

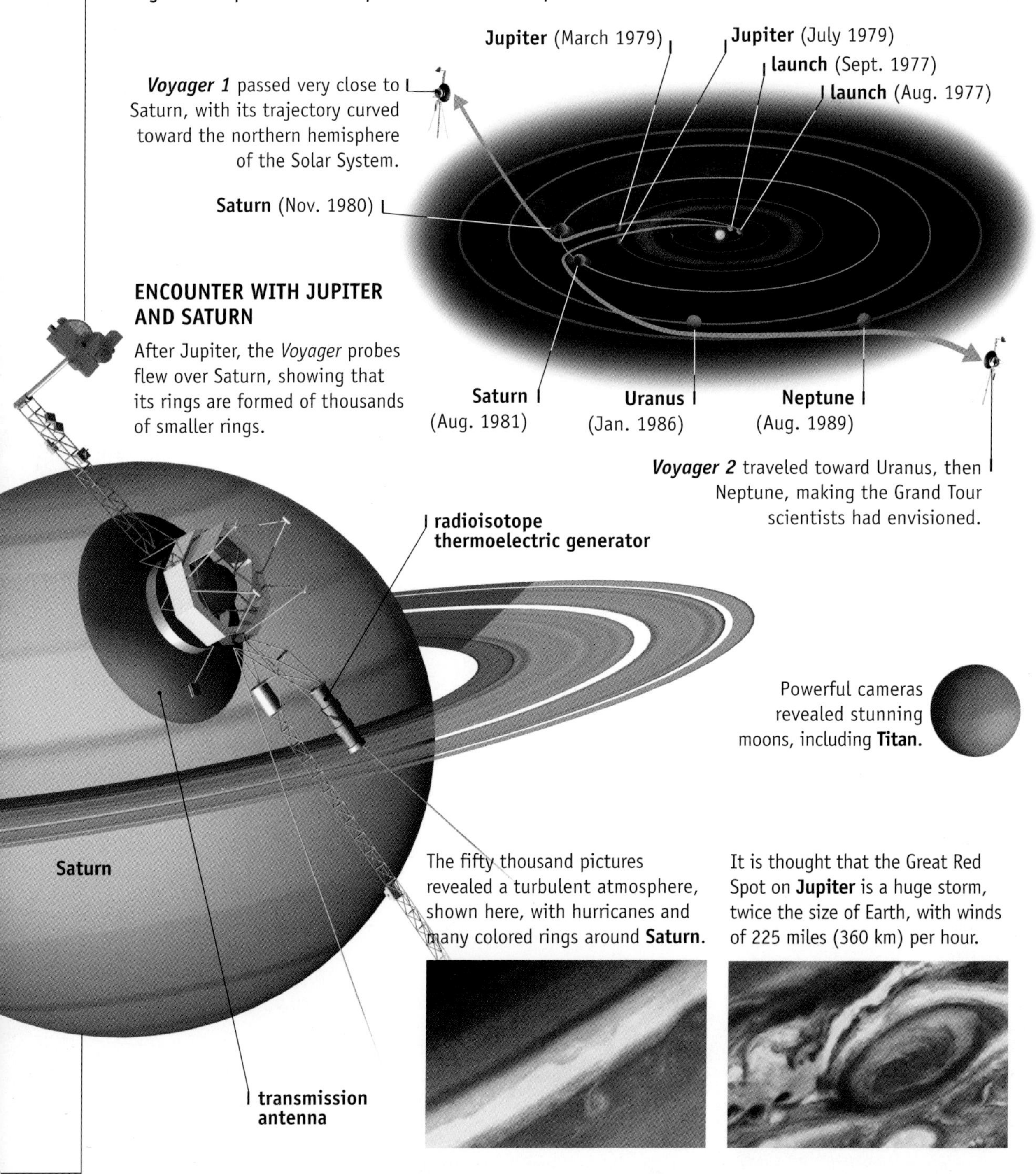

ENCOUNTER WITH JUPITER AND SATURN

After Jupiter, the *Voyager* probes flew over Saturn, showing that its rings are formed of thousands of smaller rings.

Powerful cameras revealed stunning moons, including **Titan**.

The fifty thousand pictures revealed a turbulent atmosphere, shown here, with hurricanes and many colored rings around **Saturn**.

It is thought that the Great Red Spot on **Jupiter** is a huge storm, twice the size of Earth, with winds of 225 miles (360 km) per hour.

Magellan

Unveiling the true face of Venus

Between 1960 and 1983, the Soviets sent some fifteen *Venera* probes to Venus. To gain a more complete view of the planet, the U.S. built a probe equipped with a powerful radar system. This probe, called *Magellan*, was launched in May 1989, and it entered orbit around Venus in August 1990. When its mission ended, in 1994, *Magellan* plunged into the Venusian atmosphere to test aerobraking techniques that would be used in future probes.

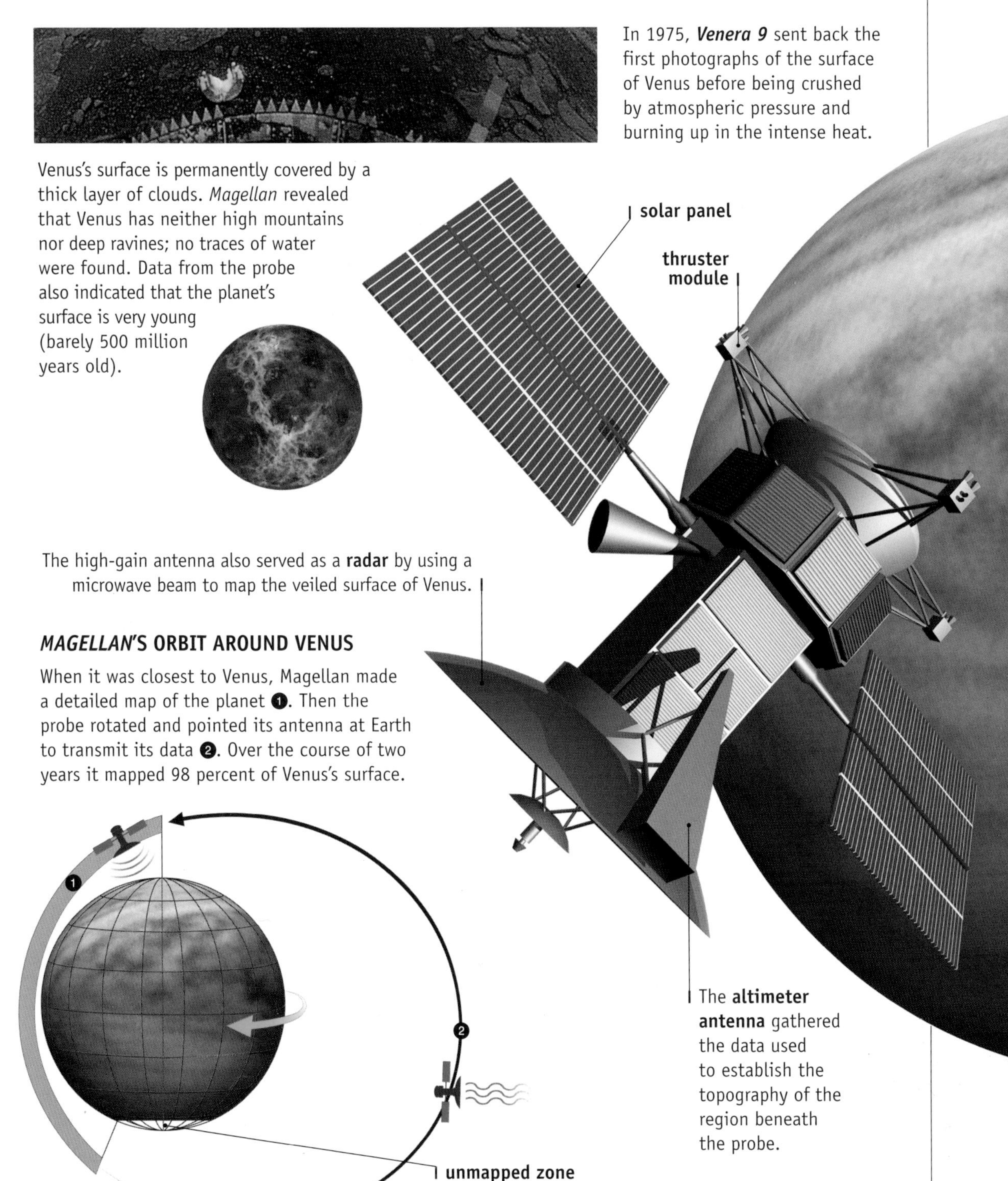

In 1975, ***Venera 9*** sent back the first photographs of the surface of Venus before being crushed by atmospheric pressure and burning up in the intense heat.

Venus's surface is permanently covered by a thick layer of clouds. *Magellan* revealed that Venus has neither high mountains nor deep ravines; no traces of water were found. Data from the probe also indicated that the planet's surface is very young (barely 500 million years old).

The high-gain antenna also served as a **radar** by using a microwave beam to map the veiled surface of Venus.

MAGELLAN'S ORBIT AROUND VENUS

When it was closest to Venus, Magellan made a detailed map of the planet ❶. Then the probe rotated and pointed its antenna at Earth to transmit its data ❷. Over the course of two years it mapped 98 percent of Venus's surface.

The **altimeter antenna** gathered the data used to establish the topography of the region beneath the probe.

Galileo

Discovering Jupiter's satellites

The *Galileo* probe was named for the great Italian astronomer who in 1610 discovered that four moons orbit Jupiter. Launched in October 1989, it became the first probe to enter orbit around Jupiter (in December 1995). *Galileo* confirmed that Jupiter's four largest satellites have a thin atmosphere.

STUNNING IMAGES

Since it went into orbit, the probe has regularly been flying by the four giant moons discovered by Galileo and has transmitted excellent pictures of them.

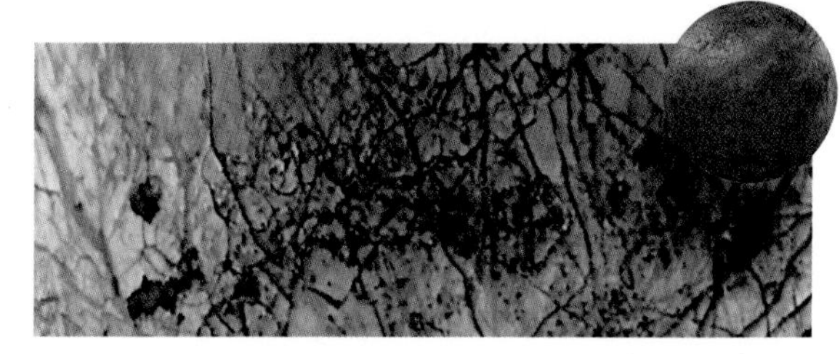

The surface of **Europa** is covered with large ice ridges running hundreds of miles (hundreds of km). Oceans of water might be found under this icy surface.

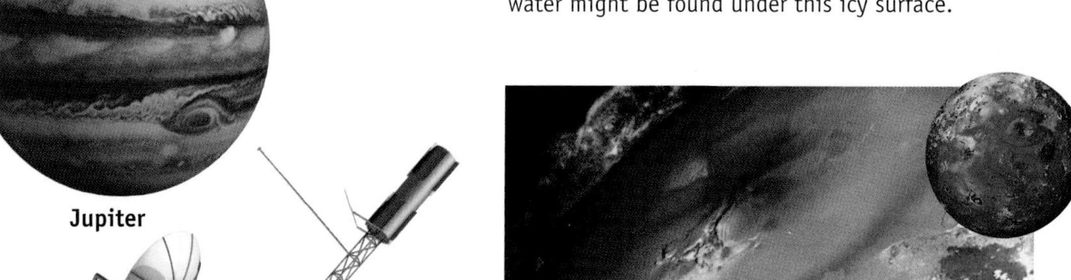

Jupiter

Io owes its stunning red, yellow, white, and orange surface to active volcanoes spewing molten sulfur.

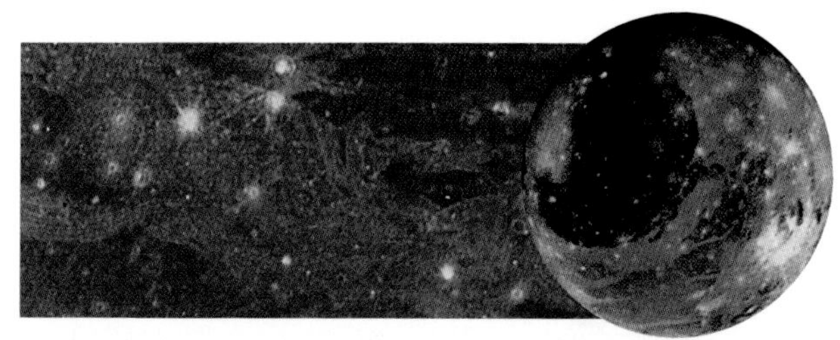

Ganymede, the largest moon in the Solar System, has a furrowed, icy surface.

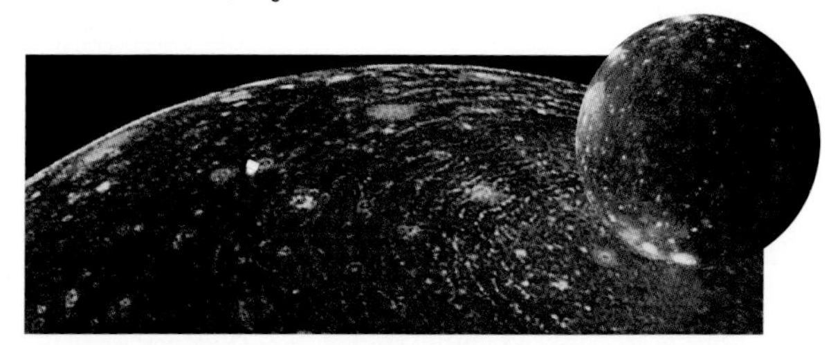

Callisto's surface is among the oldest, with the greatest number of craters, in the entire Solar System.

The probe's high-gain **antenna**, capable of transmitting billions of bits of information, unfortunately did not deploy properly.

Galileo nevertheless sends photographs in small bits, as a result of its small, low-gain antenna.

Europa

Cassini and *Huygens*

Into the mysteries of Saturn and Titan

The *Cassini* probe, launched in October 1997, will take seven years to reach Saturn, where it will orbit for four years examining the planet and its eighteen moons. *Cassini* carries instruments and a small probe, *Huygens*, which it will drop into Titan's atmosphere. The probes are named for astronomers Jean-Dominique Cassini and Christian Huygens, who made basic observations of Saturn and Titan in the seventeenth century.

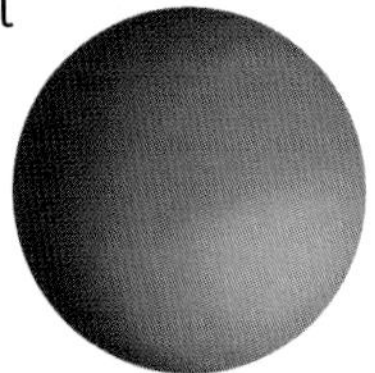

Titan

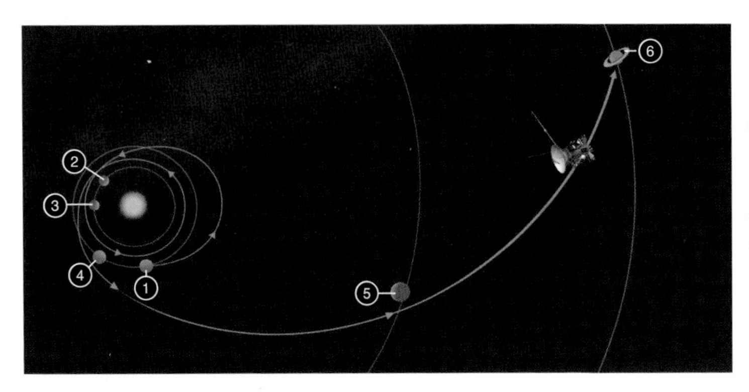

Launched in October 1997 ❶, the probes flew past Venus in April 1998 ❷ and June 1999 ❸, then Earth in August 1999 ❹. These "gravitational assistance" maneuvers were designed to increase the probes' speed for their trip to Saturn. They flew past Jupiter in December 2000 ❺ and will reach Saturn in July 2004 ❻.

If all proceeds as planned, *Cassini* will pass over Saturn's rings in July 2004, where it will fire up its main engine to brake its course and keep from being captured by the planet. It will drop *Huygens* into Titan's atmosphere in November 2004.

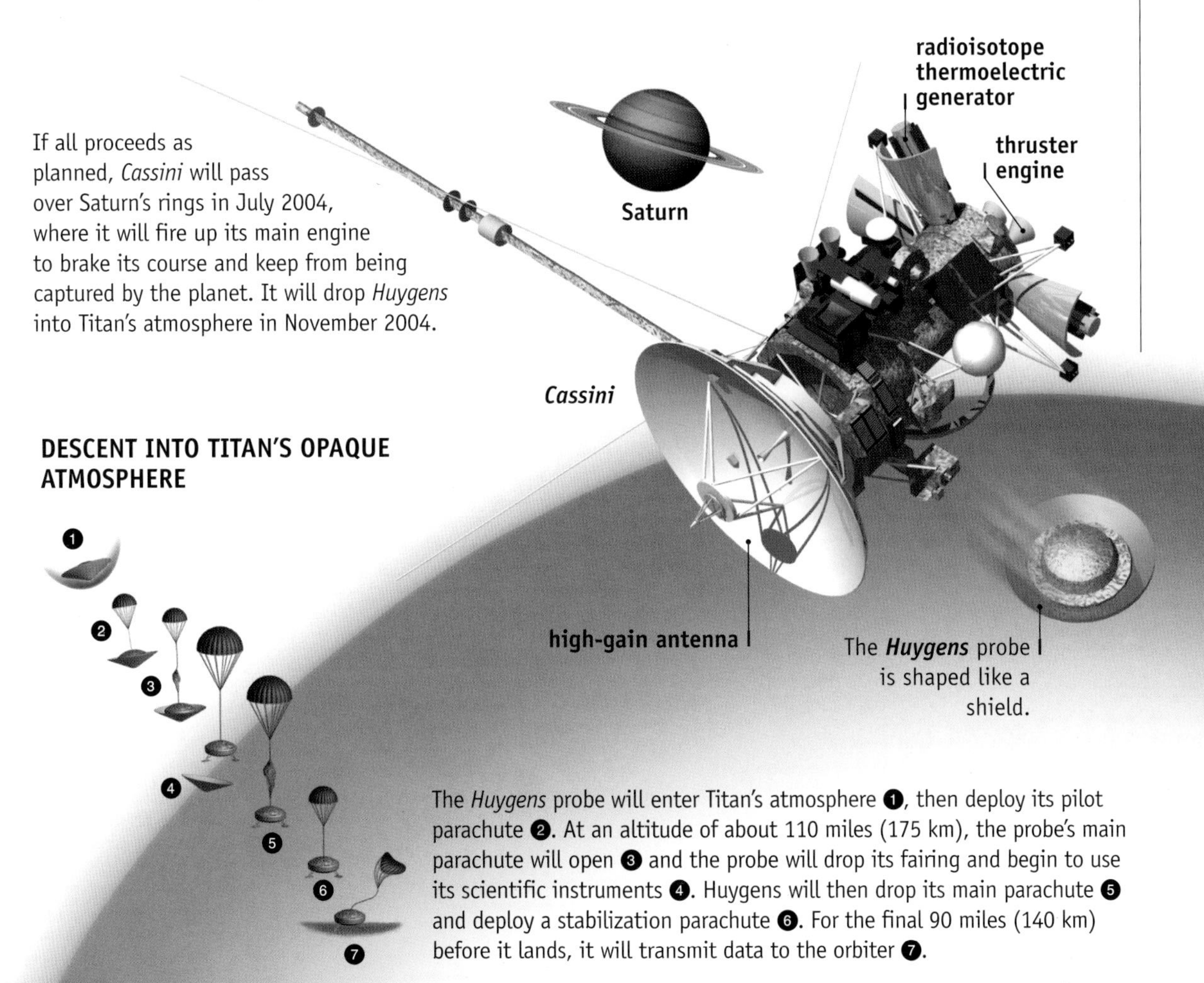

The ***Huygens*** probe is shaped like a shield.

DESCENT INTO TITAN'S OPAQUE ATMOSPHERE

The *Huygens* probe will enter Titan's atmosphere ❶, then deploy its pilot parachute ❷. At an altitude of about 110 miles (175 km), the probe's main parachute will open ❸ and the probe will drop its fairing and begin to use its scientific instruments ❹. Huygens will then drop its main parachute ❺ and deploy a stabilization parachute ❻. For the final 90 miles (140 km) before it lands, it will transmit data to the orbiter ❼.

Ulysses

Seeing the Sun's poles

Since the 1960s, dozens of satellites and probes have studied the Sun. All of them have observed our star from its equator, however, from the same perspective as we see it from Earth. To date, only one probe has been able to observe the Sun from the angle of its poles: *Ulysses*, a European probe launched in October 1990.

Since 1994, *Ulysses* has been in polar orbit around the Sun, on a six-year mission. It has observed that the solar wind blows twice as intensely at the poles as at the equator. In 2000–2001, the probe will pass the south and north poles at a time when the Sun is particularly active.

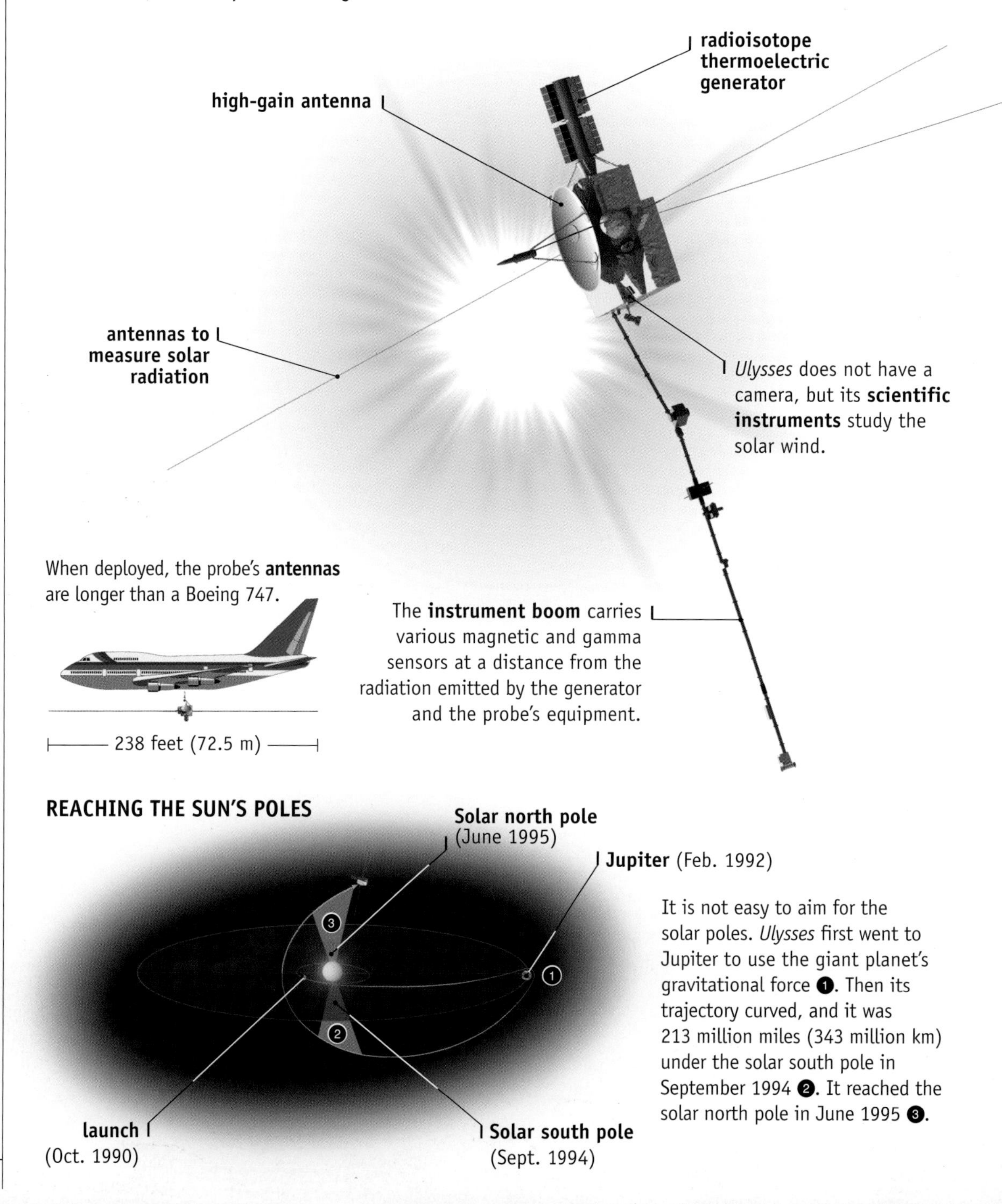

It is not easy to aim for the solar poles. *Ulysses* first went to Jupiter to use the giant planet's gravitational force ❶. Then its trajectory curved, and it was 213 million miles (343 million km) under the solar south pole in September 1994 ❷. It reached the solar north pole in June 1995 ❸.

Pathfinder

A little robot moves on Mars

More than twenty years after the *Viking* explorations, a new type of machine touched down on a Martian desert in July 1997, after a seven-month voyage. It was *Pathfinder*, a probe carrying *Sojourner*, a small all-terrain vehicle about the size of a toy truck. The probes worked until contact was abruptly lost in September 1997. Up to that point, though, the instruments collected more information than scientists had hoped for.

LANDING

Pathfinder entered the Martian atmosphere at a speed of 4.6 miles (7.4 km) per second ❶. At 7 miles (11 km) from the surface the parachute opened ❷. The back shield separated from the module ❸. The protective balloons inflated, and the retro rockets were fired ❹. After bouncing more than 15 times ❺, the probe came to rest; the balloons were deflated and retracted, and the base's petals were deployed ❻.

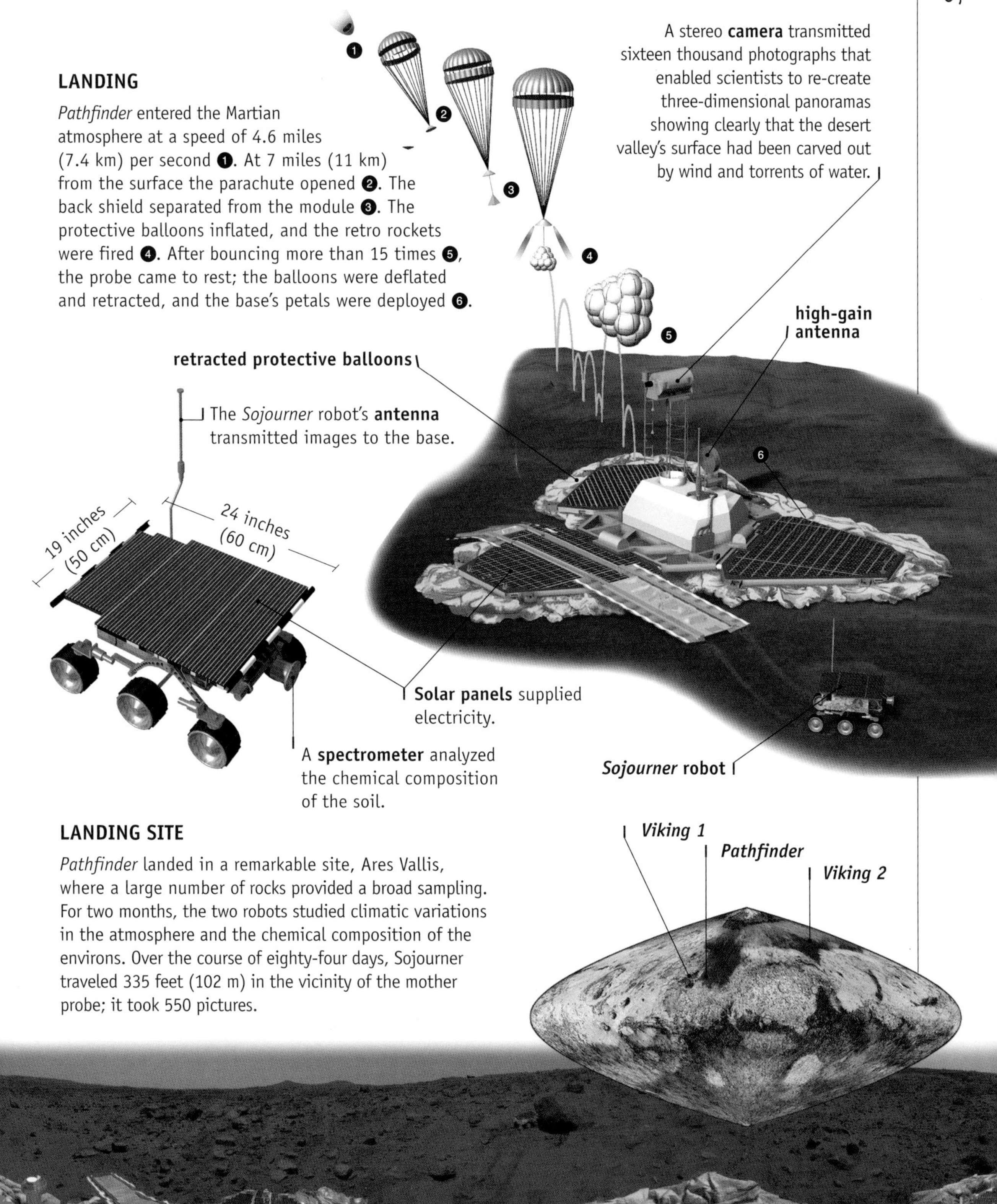

LANDING SITE

Pathfinder landed in a remarkable site, Ares Vallis, where a large number of rocks provided a broad sampling. For two months, the two robots studied climatic variations in the atmosphere and the chemical composition of the environs. Over the course of eighty-four days, Sojourner traveled 335 feet (102 m) in the vicinity of the mother probe; it took 550 pictures.

Mars Global Surveyor

Returning to Martian orbit

In September 1997, while *Pathfinder* and *Sojourner* were exploring the surface of Mars, another probe was placed in orbit around the planet. As its name indicates, the *Mars Global Surveyor* will make a general portrait of Mars. *Mars Global Surveyor* will observe climatic changes on the surface over a complete Martian year, which is equivalent to two Earth years.

The probe was to make its orbit circular by aerobraking, a technique that uses atmospheric friction to change its trajectory. One of the solar panels did not deploy properly, so operations are getting under way gradually; *Mars Global Surveyor* was in position in the spring of 1999, one year later than planned.

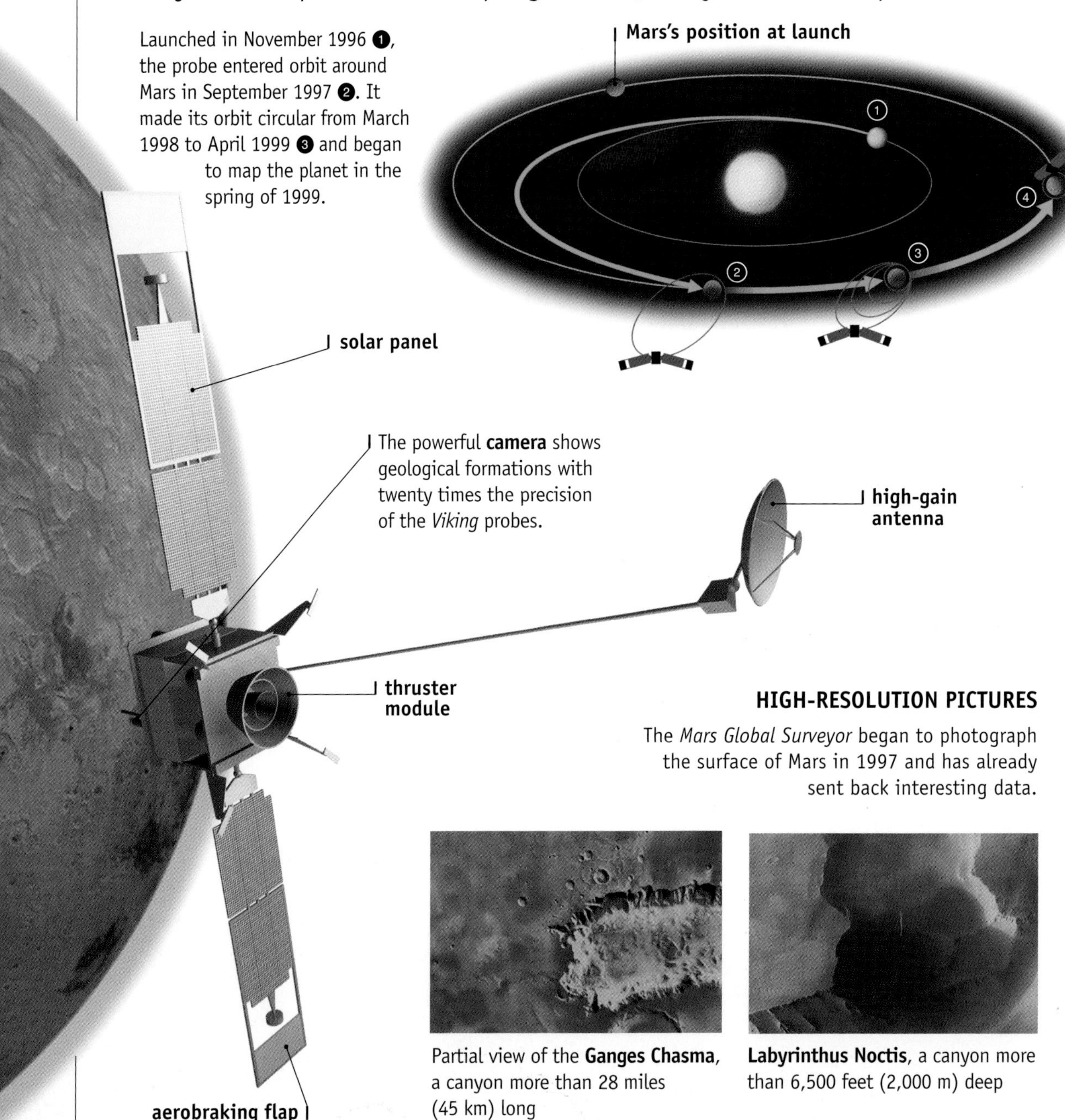

HIGH-RESOLUTION PICTURES

The *Mars Global Surveyor* began to photograph the surface of Mars in 1997 and has already sent back interesting data.

Partial view of the **Ganges Chasma**, a canyon more than 28 miles (45 km) long

Labyrinthus Noctis, a canyon more than 6,500 feet (2,000 m) deep

Clementine and Lunar Prospector

Waiting to return to the Moon

Following the "one small step for man" in the 1960s, exploration of the Moon was somewhat neglected. It was only in January 1994 that a U.S. probe, *Clementine*, once again examined our natural satellite, followed in 1998 by *Lunar Prospector*.

WATER ON THE MOON

Lunar Prospector strongly indicated that the lunar poles contain great quantities of water in the form of ice particles mixed with dust and rock. Between 33 and 985 million cubic feet (between 10 and 300 million cubic m) of water lie at the bottom of craters, covering tens of thousands of square miles (tens of thousands of sq km). There appears to be twice as much at the north pole as at the south pole. The probe is not sending back pictures of the lunar surface, but its scientific instruments are gathering data on the composition of the soil.

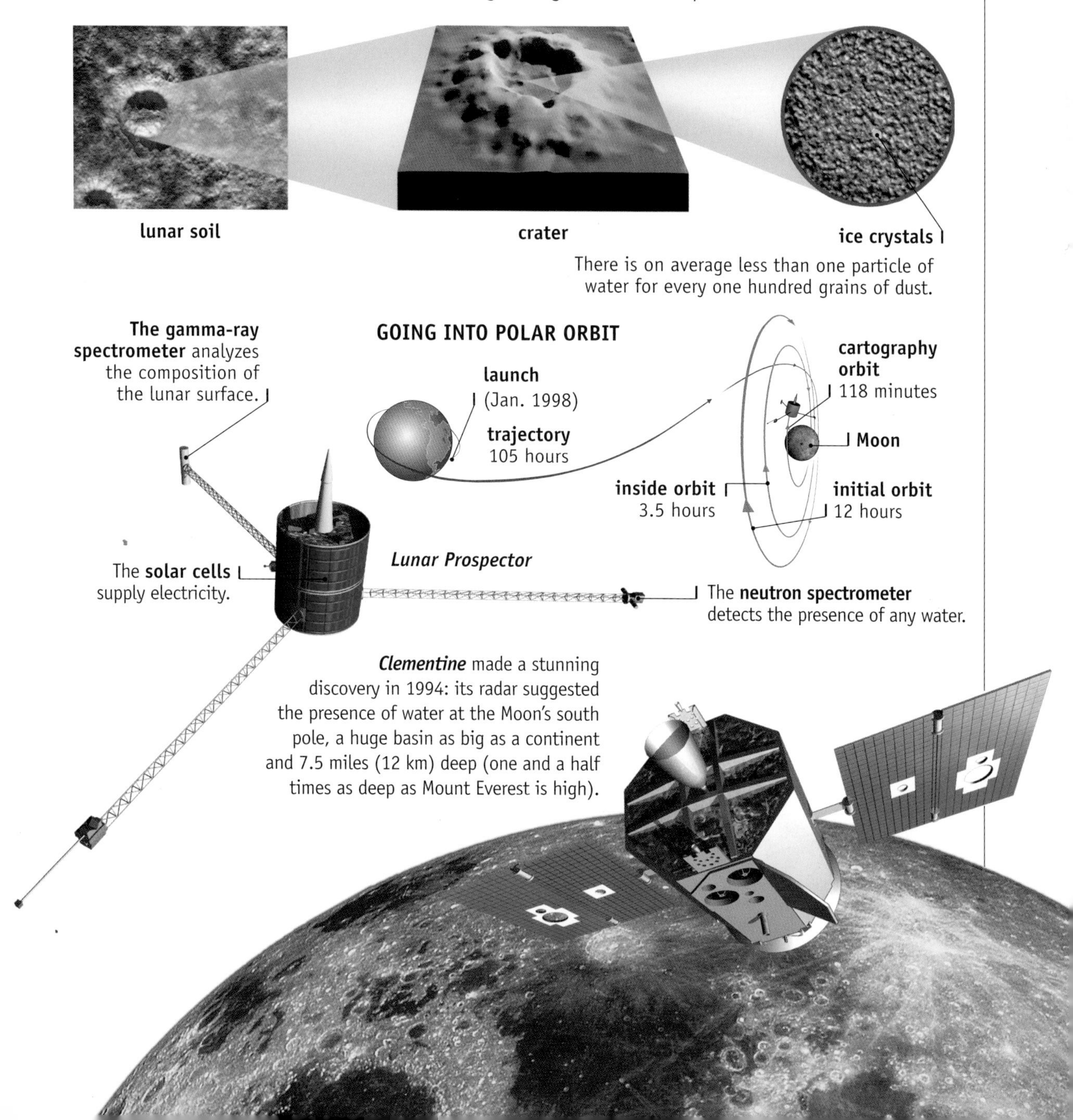

Exploring Mini-Planets

Discovering comets and asteroids

Interest in the mini-planets developed during the 1980s, when it was determined that dinosaurs probably became extinct due to one of these small heavenly bodies hitting Earth about 65 million years ago.

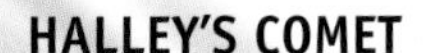

HALLEY'S COMET

In 1910, the legendary Halley's comet returned to our environs, as it does every seventy-six years, and it drew much interest. In March 1986, when Halley was circling the Sun, the *Giotto* and *Sakigake* probes went to meet it.

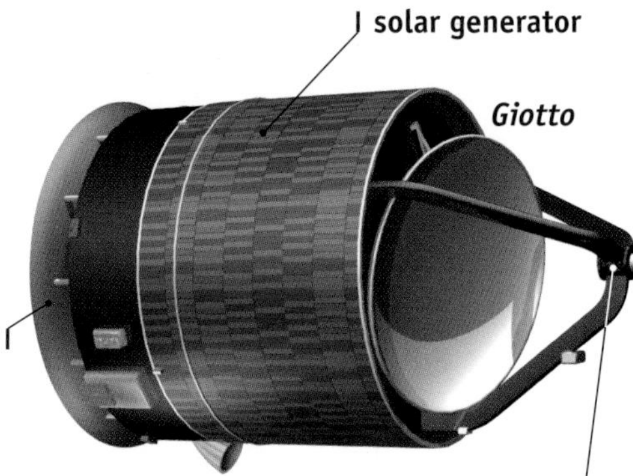

A **shield** protected the probe from dust and particles.

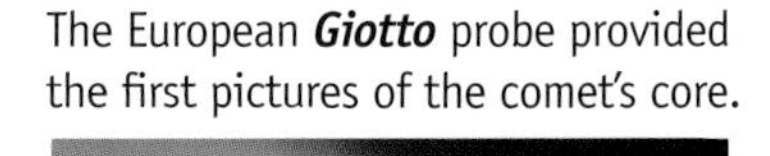

The European ***Giotto*** probe provided the first pictures of the comet's core.

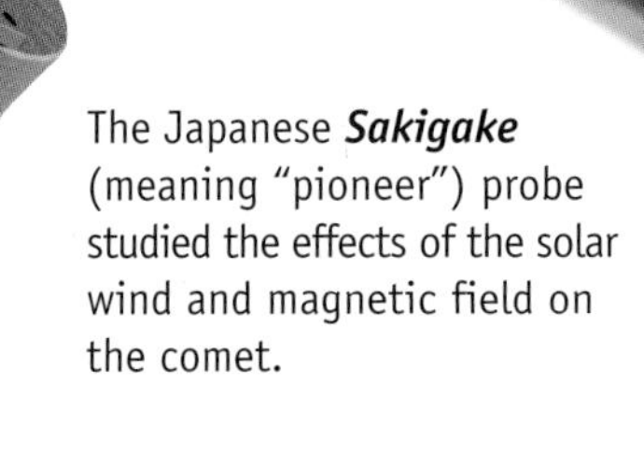

The Japanese ***Sakigake*** (meaning "pioneer") probe studied the effects of the solar wind and magnetic field on the comet.

Halley's comet

EROS AND MATHILDE: THE ASTEROIDS

Then scientists looked at asteroids, large rocks that orbit the Sun; it is feared that one could hit Earth. The *NEAR* (*Near Earth Asteroid Rendezvous*) *-Shoemaker* probe was launched by NASA in February 1996.

As it headed for Eros, *NEAR-Shoemaker* provided a good view of the asteroid **Mathilde**.

solar panels

scientific-instrument module

In February 2000, *NEAR* began orbiting **Eros** to study the asteroid closely. The mission ended with the first landing on an asteroid on February 12, 2001.

Objective: Mars

When will humans get there?

As a follow-up to the *Pathfinder* and *Mars Global Surveyor* probes, NASA plans to send probes every twenty-six months, whenever Mars's position in relation to Earth makes it possible. The ultimate objective of Mars exploration is to land human beings on the planet, which will not take place before 2018 and most likely in the 2030s.

MARS SURVEYOR 98 PROJECT

As part of the Mars Surveyor 98 project, two probes reached Mars in the fall of 1999: the *Mars Climate Orbiter* and the *Mars Polar Lander*. Unfortunately, the orbiter burned up as it entered the Martian atmosphere because of a navigation error, while the lander suddenly fell out of communication with Earth.

The lander had a **robot arm** designed to collect various samples for on-board analysis.

The ***Mars Polar Lander*** was supposed to land near the cap of the south pole, where the surface is a mixture of dust and ice.

The **orbiter**'s missions included studying climatic changes for a Martian year.

The **instrument module** was equipped with a sensor specially designed to reveal the presence of water in the atmosphere.

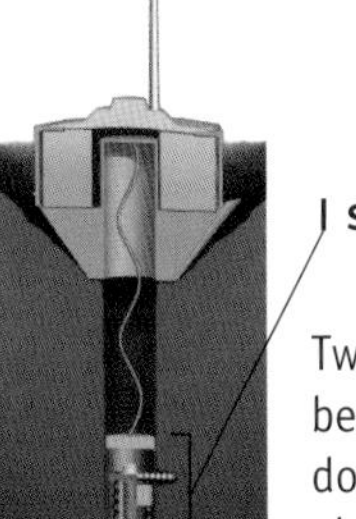

soil sampler chamber

Two **microprobes**, released just before *Mars Polar Lander* touched down, were supposed to penetrate about 3 feet (1 m) below the surface of the planet to analyze samples.

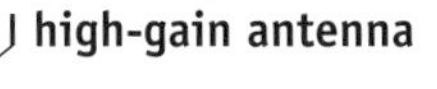

high-gain antenna

MARS ODYSSEY

In 2001, NASA plans to place into Martian orbit a probe that will analyze the chemical composition of the surface; specifically, *Mars Odyssey* will try to find possible subterranean water reserves on the Red Planet.

The gamma-ray **spectrometer** will analyze the composition of Mars's surface and try to detect the presence of hydrogen under the surface.

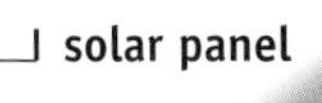

solar panel

scientific-instrument module

The Space Shuttle

An airplane for space

The Space Shuttle is the only space vehicle whose components are reused, except for the external tank (rockets, for instance, are used only once). In almost twenty years, there have been more than one hundred Shuttle launches, with only one failure (*Challenger*, in 1986). The Shuttle launched the *Galileo, Magellan,* and *Ulysses* space probes, and it placed the famous Hubble Space Telescope in orbit.

When it is launched, the orbiter is attached to a huge **tank** containing the fuel that the engines burn during the first eight minutes of flight. Every second, each engine consumes 343 gallons (1,300 liters) of the liquid hydrogen and oxygen contained in this immense tank, which is 154 feet (47 m) long and 28 feet (8.4 m) in diameter.

The main part of the Shuttle, the **orbiter**, can transport about 12 tons (11.5 metric tons) of material and five to seven astronauts into Earth orbit. It is about the size and weight of a DC-10 jet: 121 feet (37 m) long by 79 feet (24 m) wide, weighing 68 tons (61.6 metric tons) when empty.

Two solid **rocket boosters** supply most of the thrust in the first minutes of flight. The rockets (149 feet [45.5 m] tall, 12 feet [3.7 m] in diameter, and weighing 585 tons [530 metric tons]) are then jettisoned; they drop into the ocean, where they are recovered and prepared for another launch.

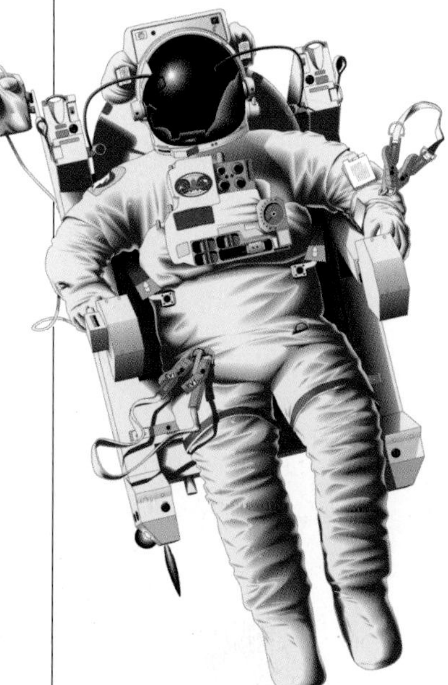

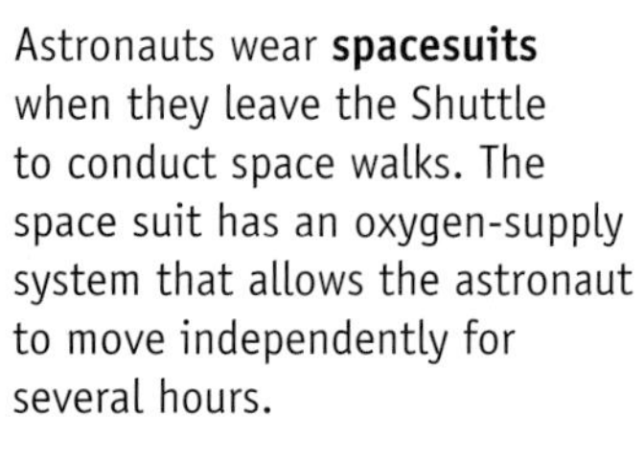

Astronauts wear **spacesuits** when they leave the Shuttle to conduct space walks. The space suit has an oxygen-supply system that allows the astronaut to move independently for several hours.

Tiles designed to resist temperatures of more than 2,300° F (1,260° C) cover 70 percent of the orbiter's surface. There are more than thirty thousand of them.

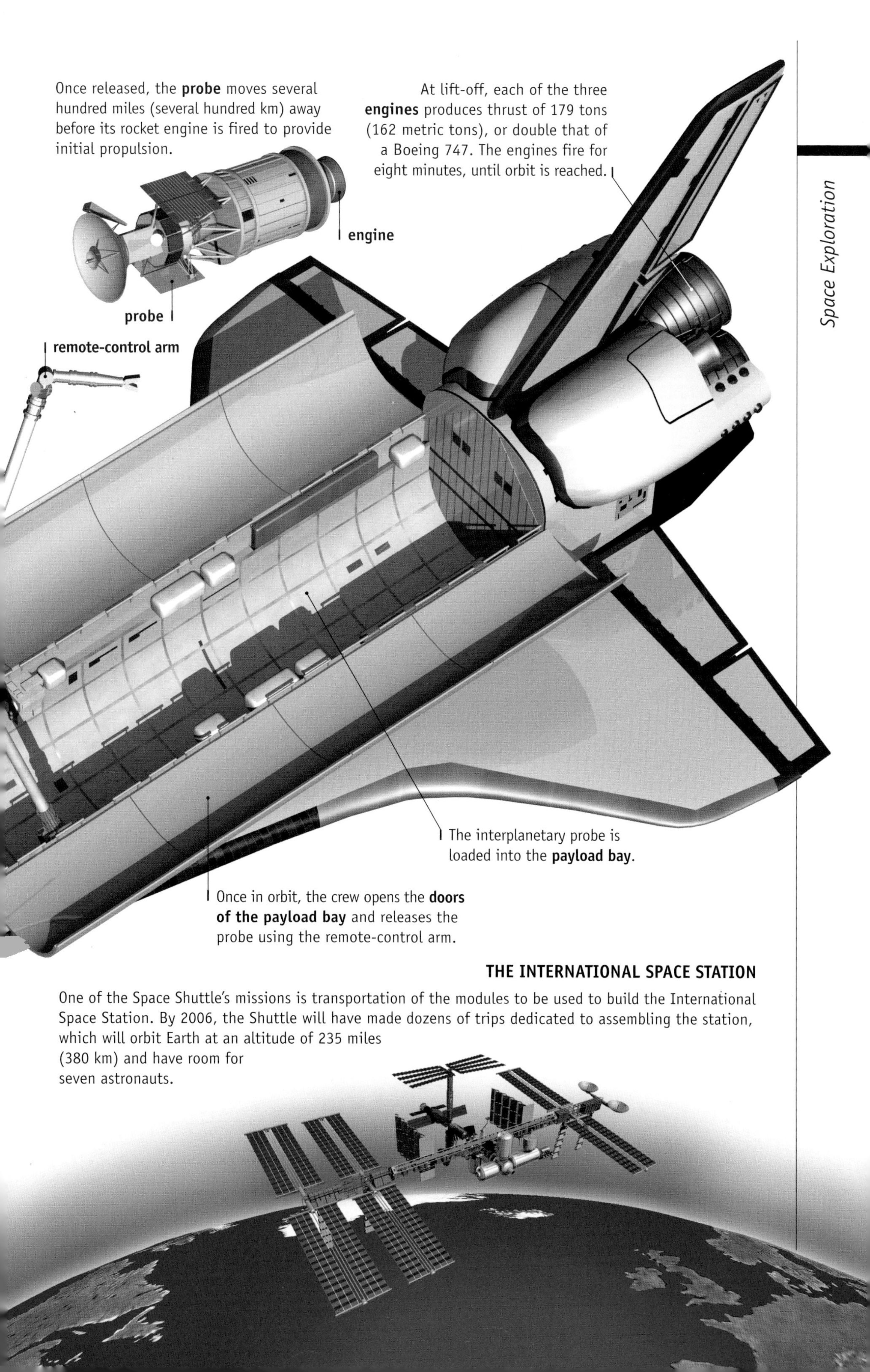

THE INTERNATIONAL SPACE STATION

One of the Space Shuttle's missions is transportation of the modules to be used to build the International Space Station. By 2006, the Shuttle will have made dozens of trips dedicated to assembling the station, which will orbit Earth at an altitude of 235 miles (380 km) and have room for seven astronauts.

Glossary

absolute zero: The lowest possible temperature; the point at which energetic molecular movement stops; it is equivalent to 0 K, -459.69° F, or -273.15° C.

albedo: The smallest unit of a chemical element that retains all of the element's characteristics.

aphelion: For an object that orbits the Sun, the point on its orbit that is farthest from the Sun.

astronomical unit (AU): Unit used to calculate distances in the Solar System. One astronomical unit corresponds to the average distance between Earth and the Sun, about 93 million miles (150 million km).

atom: The smallest unit of a chemical element that retains all of the element's characteristics. An atom is made of a nucleus (which itself is made of protons and neutrons) around which electrons orbit.

Big Bang: According to the theory of many scientists, the cosmic explosion that marked the origin of the Universe.

black hole: A small celestial body with an intense gravitational field so strong that not even light can escape being drawn in. Black holes are believed to be collapsed stars.

electromagnetic radiation: Energy transmitted at the speed of light in the form of gamma rays, X rays, ultraviolet rays, visible light, infrared waves, and radio waves.

electromagnetic spectrum: The complete range of electromagnetic radiation, extending from gamma rays (the shortest wavelengths) to radio waves (the longest wavelengths).

fairing: The external surface of an aircraft or a structure that is attached to the outside of an aircraft and designed to reduce drag.

gamma rays: Very high-energy radiation with the shortest wavelength.

gravitational assistance: The technique of using a planet's gravitational field to change a probe's trajectory and increase, or boost, its speed, without the additional consumption of fuel.

hydrogen: The lightest and most abundant chemical element in the Universe. Its nucleus is composed of one proton, around which one electron orbits.

infrared waves: Electromagnetic radiation whose wavelength is slightly longer than that of visible light; heat.

interferometry: In radio astronomy, the technique of combining light beams captured by two or more radio telescopes to increase the resolution (sharpness) of images.

light-year: The distance traveled by light in one year at the speed of about 5.878 trillion miles (9.46 trillion km) per year.

NGC: Abbreviation for *New General Catalogue of Nebulae and Star Clusters*. This document is used to identify celestial objects that are not stars.

oscillation: a single swing in one direction of a body that vibrates, swings, or moves back and forth.

parsec: The unit of distance equivalent to 3.26 light-years or 206,265 AU.

perihelion: For an object that orbits the Sun, the point in its orbit that is closest to the Sun.

photon: Particle that transmits electromagnetic radiation, including visible light.

primordial: Being or happening first in the sequence of time; original.

quark: A charged elementary particle; the smallest particle of matter; a component of protons and neutrons.

quasar: a "quasi-stellar" object; any of numerous starlike objects that may be the most distant and brightest objects in the Universe.

radio waves: Part of the electromagnetic spectrum in which the wavelength varies from about four one-hundredths of an inch (0.0394 inch) (0.1 cm) to several yards (several m) or even several miles (several km). Radio radiation has the longest wavelength.

visible light: The narrow segment of the electromagnetic spectrum that is visible.

Books

Adventure in Space: The Flight to Fix the Hubble. Elaine Scott and Margaret Miller (Disney Press)

The Adventures of Sojourner: *The Mission to Mars that Thrilled the World*. Susi Trautmann Wunsch (Mikaya Press)

The Backyard Astronomer's Guide. Terence Dickinson, Alan Dyer (Cambridge House Publishing)

The Hole in the Universe: How Scientists Peered over the Edge of the Universe and Found Everything. K. C. Cole (Harcourt Brace)

Hubble Space Telescope: Exploring the Universe (Countdown to Space). Michael D. Cole (Enslow)

NASA and the Exploration of Space. Roger D. Launius (Stewart)

Nightwatch: A Practical Guide to Viewing the Universe. Terence Dickinson (Firefly Books)

One Universe : At Home in the Cosmos. Neil De Grasse Tyson, Robert Irion, and Charles Tsun-Chu Liu (Joseph Henry Press)

Space Exploration: A Pro/Con Issue (Hot Pro/Con Issues). Sarah Flowers (Enslow)

The Universe and Beyond. Terence Dickinson (Firefly Books)

Videos

Apollo 13 (Universal Studios)

Astronomy 101: A Beginner's Guide to the Night Sky (Mazon Productions, Inc.)

Life Beyond Earth (PBS Home Video)

NASA: 25 Years of Triumph and Tragedy (Madacy Records)

Nova: To the Moon (WGB)

Star Gaze (D3/Digital Disc)

The Ultimate Space Experience (Madacy Records)

The Voyager *Odyssey* (Image Entertainment)

Web Sites

Astrogirl
www.astrogrl.org/

Imagine the Universe
imagine.gsfc.nasa.gov/

NASA — Origins
origins.jpl.nasa.gov/

StarChild: Learning Center
starchild.gsfc.nasa.gov/docs/StarChild/StarChild.html

Index

Index